BUILDING BIG IDEAS in Math

With a Frog Theme

Grades 6 - 8

Evan Maletsky
Montclair State University

Mary Kay Varley
Fort Worth Country Day School

Published by Didax, Inc., Rowley, MA USA

Printed in the United States of America.

Order Number 2-157
ISBN 1-58324-192-2

A B C D E F G 08 07 06 05 04

395 Main Street
Rowley, MA 01969
www.didax.com

Table of Contents

Introduction iv

Getting Started v

1. Patterns in Arrays .... FINDING PATTERNS 1
Counting and Building with Triangular Arrays
BIG IDEA: Triangular Numbers

2. Patterns in Letters .... ARRANGING LETTERS 17
Arranging Letters and Counting Choices
BIG IDEA: Factorials

3. Patterns in Moves .... PLANNING HOPS AND JUMPS 31
Playing by the Rules
BIG IDEA: Thinking Strategies

4. Patterns in Choices .... COMPUTING CHANCES 43
Using Tree Diagrams to Find Probabilities
BIG IDEA: Tree Diagrams

5. Patterns in Pictures .... DRAWING PICTURES 55
Listing Arrangements as Pictures
BIG IDEA: Networks

6. Patterns in Designs .... TRAVELING ROUTES 67
Testing for Traceability
BIG IDEA: One-Line Drawings

7. Patterns in Turns .... REPEATING STEPS 81
Drawing Repeating Patterns
BIG IDEA: Iteration

8. Patterns in Positions .... PUTTING IT TOGETHER 95
Seeing Slides and Turns
BIG IDEA: Translations and Rotations

Puzzles for Problem Solving 109

Writing and Reading Ideas 120

Introduction

Mathematics must tickle the senses as well as stretch the mind.

Mathematics is much more than just a collected set of computational skills that are reviewed repeatedly through practice and drill. It includes the recognition and extension of patterns in numbers and pictures as well as the ability to create and solve problems and puzzles. Furthermore, it should be exciting and challenging, visual and dynamic, animated and appealing, and provoke not only student interest but student curiosity as well. The mathematical experiences of the classroom should build on the familiar, but they should also include new and different ideas and connections that make students see and reach beyond the familiar. It is with these thoughts in mind that the authors have created this collection of activities built around the theme of frogs.

The math activities in this book are designed for classroom use at the middle grades, 6 through 8. They are all hands-on experiences, readily accessible and adaptable to a wide variety of abilities and backgrounds. Each activity includes a specific problem to be solved, with extensive teacher notes and appropriate mathematical background, a complete solution, plus suggested extensions, student worksheets, and transparency masters. These activities have been tested extensively and contain many teaching suggestions and actual student reactions reproduced just as they were written by the students.

All the activities support and supplement key components of basic existing curricula and reflect the recommendations of national and state standards in mathematics. From a mathematical point of view, the content emerges from the fundamental notions of discrete mathematics. Discrete mathematics has become an increasingly important component of the mathematics curriculum and is now cited in many state standards. It deals with a wide variety of topics and techniques that center around arranging, ordering, counting, and connecting discrete objects. Many familiar activities involving pattern recognition fall into this category as do an increasingly wide variety of important applications in everyday life.

Enjoy these activities.
Enjoy the frogs.
Use them both to help your students enjoy mathematics.

Evan Maletsky
Montclair State University
Upper Montclair, NJ

Mary Kay Varley
Fort Worth Country Day School
Fort Worth, TX

Getting Started

This book builds big ideas in mathematics in grades 6 through 8 through innovative, hands-on classroom activities. These activities are designed

- to motivate interest and enjoyment in mathematics,
- to bring together different mathematical topics under the single theme of frogs,
- to give students a broad and encompassing view of mathematics,
- to emphasize the importance of recognizing, extending, and applying patterns,
- to utilize simple, easily made manipulatives, and
- to be adaptable to a range of ages, abilities, and backgrounds.

Each section is made up of five key components.

Introduction - an opening page description of the activity, expected student outcomes, the **BIG IDEA** (marked by the icon at left), as well as connections to the NCTM Standards

Problem - a two-page spread that walks you through the steps needed prior to the presentation of a single key problem, plus the solution and possible extensions

Worksheets - two student worksheets keyed to the topic of the initial classroom problem along with detailed answers

Transparencies - a collection of transparency masters to use in conjunction with the problem, the worksheets, and the extensions

Suggestions and Hints - extensive suggestions for teaching the activity plus student reactions, written in their own hand

Every section focuses on several of the content standards of the National Council of Teachers of Mathematics *Curriculum and Evaluation Standards*. These specific connections are noted on the bottom right of each opening page. Use the empty space on the left to write in the related state standards for your particular school.

There are five NCTM process standards.

problem solving
reasoning and proof
communication
connections
representation

These standards receive special attention throughout, with specific identification and commentary. Watch for the **Math Standard** logo (at left) in the left hand margins.

You will find short **FROG Facts** throughout the booklet. These are designed to add touches of human interest and bits of science. In addition, between activities there are **MATH and FROGS** connections that can serve as the basis for additional computation, problem-solving, reading, and research.

MATH and FROGS

COUNTING

There are about 4000 different species of frogs and toads.

One of the most striking is the red-eyed tree frog, with its slender body and enormous eyes. Thousands upon thousands of these colorful tiny frogs are in the rain forests of Central America, but they would be very hard to find and count.

Introduction to Activity 

FINDING PATTERNS
Counting and Building with Triangular Arrays

Description

Many people view mathematics as the science of patterns. This activity uses a folded paper strip to create a counting situation that leads to an interesting number pattern.

1 = **1**

Students try to count mentally all the rectangles they would see in a strip of paper folded in half twice. When the strip is opened in their hands, they letter each of the parts and count rectangles visually, using letters to name the rectangles.

1 + 2 = **3**

The answer is the sum of four successive counting numbers starting with 1. When generalized, this counting process leads to the triangular numbers that can be viewed geometrically as triangular arrays. As a final counting problem, students count paths down these triangular arrays and discovery another interesting triangular number pattern.

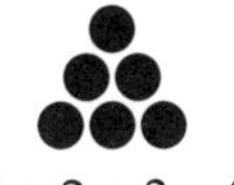

1 + 2 + 3 = **6**

Big Idea

Triangular numbers are sums of successive sets of counting numbers starting with 1. The ancient Greeks viewed these triangular numbers geometrically as triangular arrays.

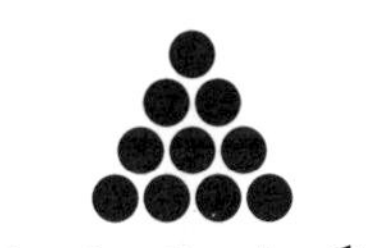

1 + 2 + 3 + 4 = **10**

Expected Outcomes

Students will gain experience in

- finding and extending patterns
- counting systematically
- building triangular arrays
- visualizing parts of geometric figures
- connecting geometry and arithmetic

1 + 2 + 3 + 4 + 5 = **15**

Meeting these NCTM Standards

✓ Numbers
✓ Algebra
✓ Geometry
Measurement
Data Analysis
✓ Problem Solving
✓ Reasoning and Proof
✓ Communication
✓ Connections
✓ Representation

Activity 1 FINDING PATTERNS

Getting Ready

Begin by reviewing the definition of a rectangle.

Give each student a 2 × 8-inch strip of paper to hold in their hands. Have them fold the paper strip in half as shown.

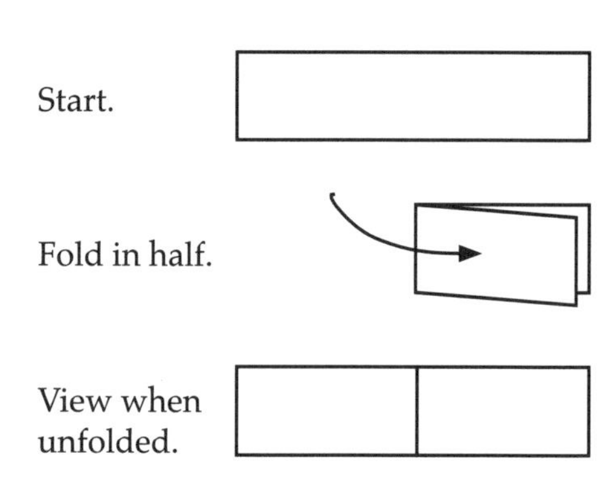

> **Ask:** *Without opening the strip, how many rectangles do you think you will see when the strip is unfolded?*
>
> Their answer should be 3, the original one plus two smaller ones.

Let the students check their answers by unfolding the paper strip and counting what they see. Then lead them into this **problem-solving** situation.

Problem

Refold the paper strip and then fold in half a second time.

Fold in half.

Fold in half again.

View when unfolded.

> **Ask:** *Without opening the strip, how many rectangles do you think you will see when the strip is unfolded?*
>
> There are 10 rectangles in all, including the 4 single squares that are also rectangles.

Seeing the Solution

To help students see the 10 rectangles, have them label the four squares with the letters F, R, O, and G. Then use the letter names as **representations** of the different rectangles.

FROG	FRO	FR	F
	ROG	RO	R
		OG	O
			G

There is 1 name with four letters, 2 with three, 3 with two, and 4 with one letter. Together, the total sum is 10.

$$1 + 2 + 3 + 4 = 10$$

Connections

The sums of successive counting numbers form a special set of numbers called **triangular numbers**.

1 3 6 10 15 21 28 . . .

$$1 = \mathbf{1}$$
$$1 + 2 = \mathbf{3}$$
$$1 + 2 + 3 = \mathbf{6}$$
$$1 + 2 + 3 + 4 = \mathbf{10}$$
$$1 + 2 + 3 + 4 + 5 = \mathbf{15}$$
$$1 + 2 + 3 + 4 + 5 + 6 = \mathbf{21}$$
$$1 + 2 + 3 + 4 + 5 + 6 + 7 = \mathbf{28}$$

These diagrams will help you see why this set of numbers is called the triangular numbers.

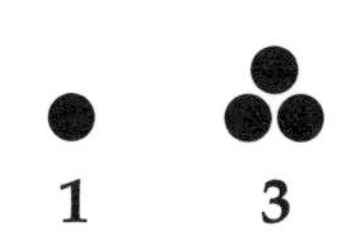
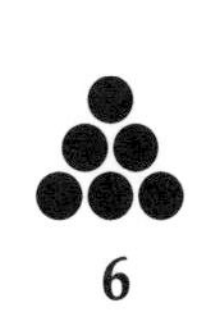

1 **3** **6** **10** **15**

With a paper strip folded into 4 parts, the total number of rectangles is 10, the fourth triangular number. This leads into an interesting generalization.

Extension

Show the **connection** between the triangular numbers and the number of rectangles when the strip is folded into any number of different parts.

Parts	1	2	3	4	5	6
Rectangles	1	3	6	10	15	21

1 part

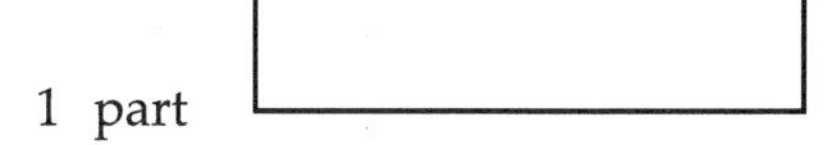

2 parts

3 parts

4 parts

5 parts

The answers are always triangular numbers. For n parts, the answer is the nth triangular number.

$$n\text{th triangular number} = \frac{n(n+1)}{2}$$

As an interesting variation, have your students count all possible paths down some triangular arrays. See how many ways they can spell out the words.

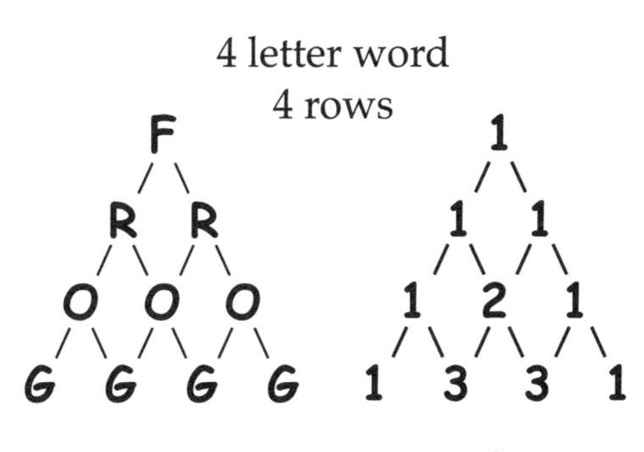

1 + 3 + 3 + 1 = 8 paths

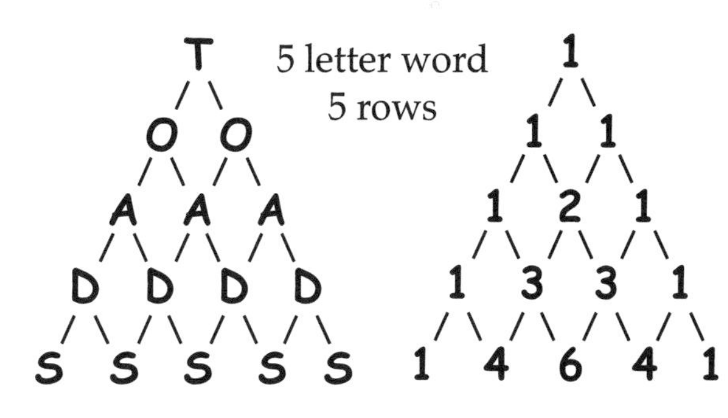

1 + 4 + 6 + 4 + 1 = 16 paths

COUNTING RECTANGLES Worksheet 1A

Start with a 2 x 8-inch strip of paper in the form of a rectangle.

1. Fold the paper strip in half and in half again, as shown. Keep it folded in your hand. Try to imagine in your mind what it will look like when unfolded, and count all the rectangles that you think you will see.

____

2. Now open the strip. Look with your eyes at the figure and count again all the rectangles that you see. Don't forget to count the squares. Squares are rectangles, special rectangles with all sides the same length.

____

3 . Print the four letters of the word FROG in the four squares of the opened strip. Count the different rectangles again. Only this time use the appropriate letters to name the rectangles as you find them. Make a complete list. Then check to see if you have the same number of rectangles that you found above.

____

4. How many rectangles can you name in each of these figures?

____ F R O

____ F R

____ F

TRIANGULAR ARRAYS

Worksheet 1B

1. Count the squares in each array. Enter the numbers in the table. Look for a pattern. Then see if you can extend the pattern for a triangular array with 5 rows.

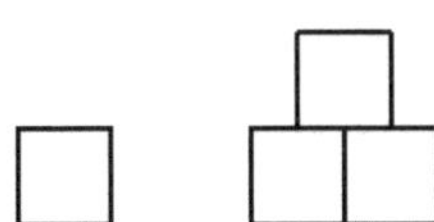

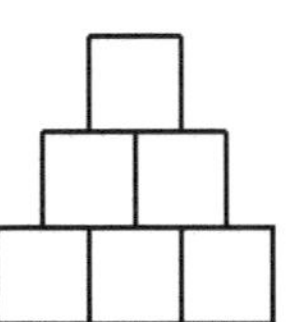

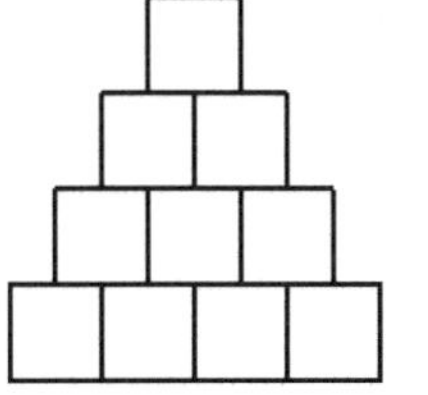

Number of rows	1	2	3	4	5
Number of squares					

The numbers of squares in these triangular arrays are called **triangular numbers**. They are the sums of successive counting numbers that start with 1.

2. Find these triangular numbers written as sums of successive counting numbers.

3. Find the seventh, eighth, ninth, and tenth triangular numbers.

$1 =$ ______

$1 + 2 =$ ______

$1 + 2 + 3 =$ ______

$1 + 2 + 3 + 4 =$ ______

$1 + 2 + 3 + 4 + 5 =$ ______

$1 + 2 + 3 + 4 + 5 + 6 =$ ______

______

______

______

______

Now study these triangular arrays of letters. In each case, count all the paths down from the top that spell out the word.

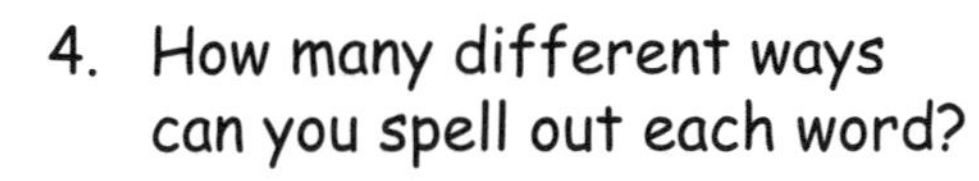

4. How many different ways can you spell out each word?

HOP ______

FROG ______

TOADS ______

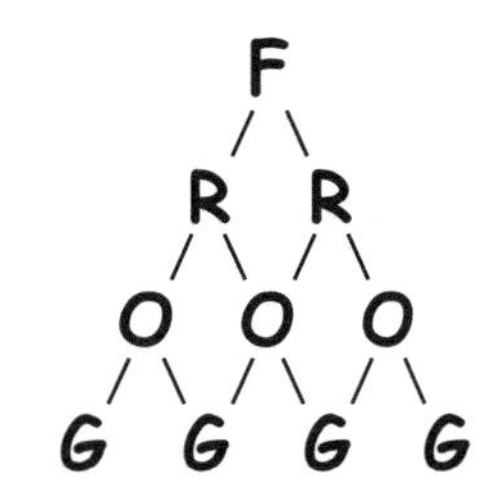

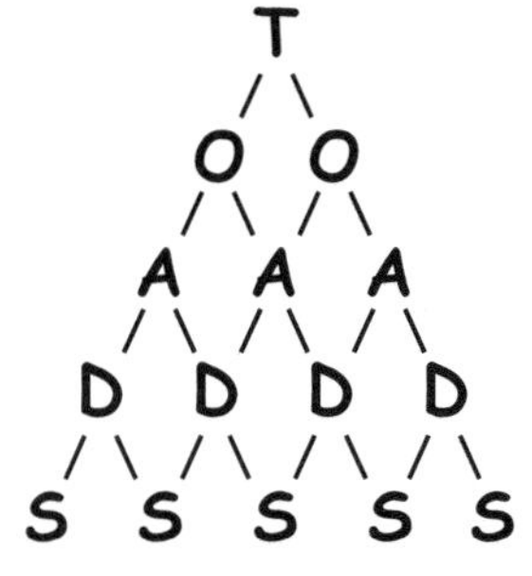

ANSWERS to Worksheets 1A and 1B

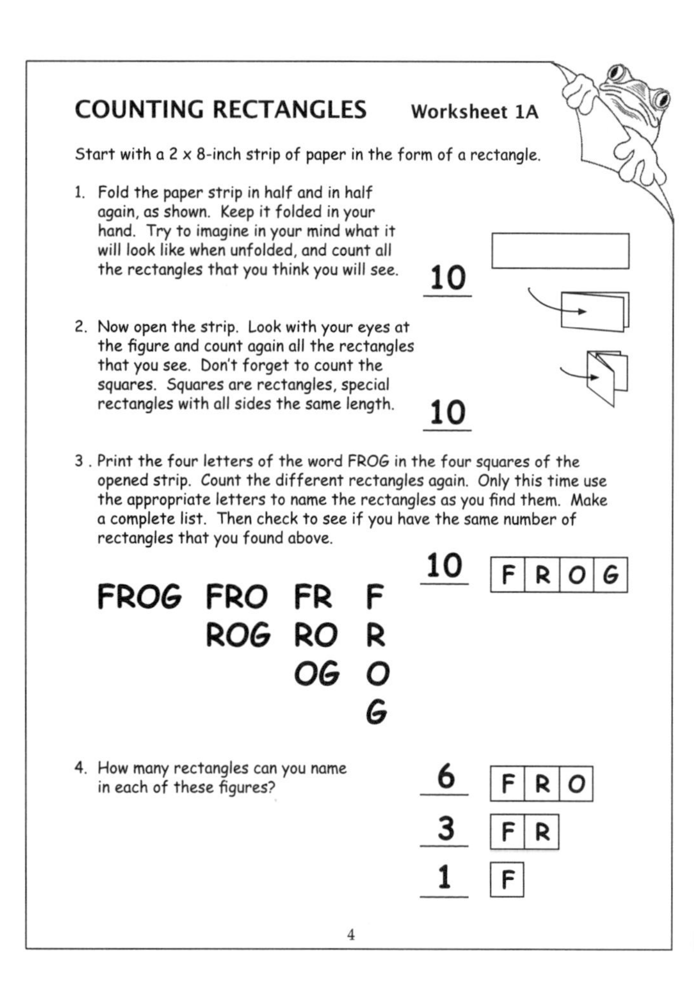

COUNTING RECTANGLES — Worksheet 1A

Start with a 2 x 8-inch strip of paper in the form of a rectangle.

1. Fold the paper strip in half and in half again, as shown. Keep it folded in your hand. Try to imagine in your mind what it will look like when unfolded, and count all the rectangles that you think you will see. **10**

2. Now open the strip. Look with your eyes at the figure and count again all the rectangles that you see. Don't forget to count the squares. Squares are rectangles, special rectangles with all sides the same length. **10**

3. Print the four letters of the word FROG in the four squares of the opened strip. Count the different rectangles again. Only this time use the appropriate letters to name the rectangles as you find them. Make a complete list. Then check to see if you have the same number of rectangles that you found above. **10**

F	R	O	G

FROG FRO FR F
ROG RO R
OG O
G

4. How many rectangles can you name in each of these figures?

6 — F R O
3 — F R
1 — F

4

TRIANGULAR ARRAYS — Worksheet 1B

1. Count the squares in each array. Enter the numbers in the table. Look for a pattern. Then see if you can extend the pattern for a triangular array with 5 rows.

Number of rows	1	2	3	4	5
Number of squares	**1**	**3**	**6**	**10**	**15**

The numbers of squares in these triangular arrays are called **triangular numbers**. They are the sums of successive counting numbers that start with 1.

2. Find these triangular numbers written as sums of successive counting numbers.

$1 =$ **1**
$1 + 2 =$ **3**
$1 + 2 + 3 =$ **6**
$1 + 2 + 3 + 4 =$ **10**
$1 + 2 + 3 + 4 + 5 =$ **15**
$1 + 2 + 3 + 4 + 5 + 6 =$ **21**

3. Find the seventh, eighth, ninth, and tenth triangular numbers.

28
36
45
55

Now study these triangular arrays of letters. In each case, count all the paths down from the top that spell out the word.

4. How many different ways can you spell out each word?

HOP **4**
FROG **8**
TOADS **16**

H
O O
P P P

F
R R
O O O
G G G G

T
O O
A A A
D D D D
S S S S S

5

FROG Facts

Herpetology is a branch of zoology that deals with reptiles and amphibians. Frogs are amphibians. So a scientist that studies frogs is called a herpetologist.

her′pe•tol′o-gist

TRANSPARENCY MASTERS

Make transparencies from these masters.

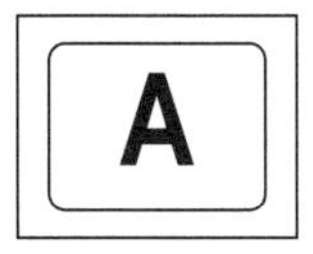

Use the left half of this transparency to show a list of the 10 different rectangles that can be found from the four connected squares lettered with the word FROG. Use the right half to extend the number pattern to include five connected squares lettered with the word FROGS.

Use this transparency to show the first ten triangular numbers as sums of successive counting numbers that start with 1. Once students see how they are being generated, ask for the quickest way to find the eleventh triangular number. See how many students suggest adding 11 to 55.

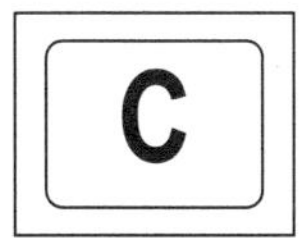

The early Greeks saw numbers as geometric figures. This transparency shows successive triangular arrays and their corresponding numerical values, just as they were seen in ancient times. Note how the first differences for the triangular numbers increase by 1 each time.

This shows the eight different paths that spell out the 4-letter word FROG, each in its own triangular grid. Note the different locations of the final letter G in the bottom row. There is 1 way to end at the far left, 1 at the far right, and 3 for the middle two locations.

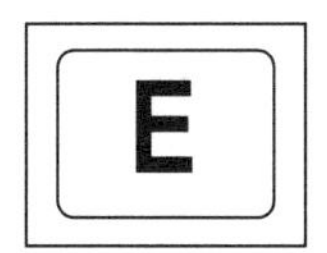

This transparency connects the 16 different paths that spell the word TOADS to the numbers in row 5 of Pascal's triangle, shown on transparency F.

With this transparency, students can also begin to discover, explore, and extend the number patterns in Pascal's triangle.

COUNTING RECTANGLES

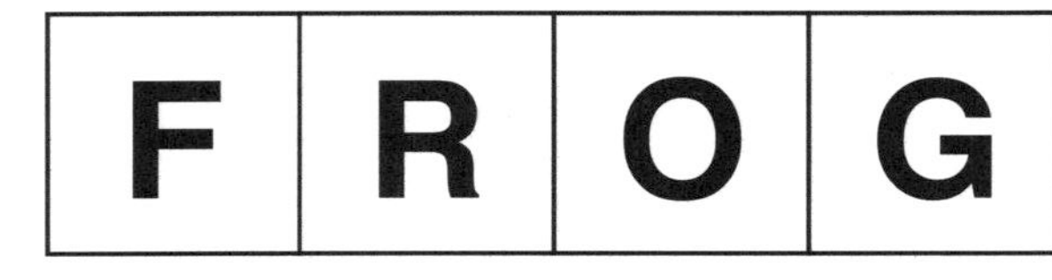

FROG	FRO	FR	F
	ROG	RO	R
		OG	O
			G

1 + 2 + 3 + 4 = 10

4 parts form
10 rectangles.

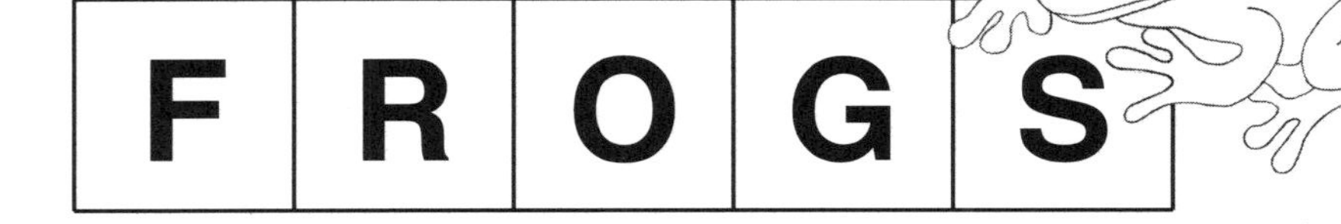

FROGS	FROG	FRO	FR	F
	ROGS	ROG	RO	R
		OGS	OG	O
			GS	G
				S

1 + 2 + 3 + 4 + 5 = 15

5 parts form
15 rectangles.

A

TRIANGULAR NUMBERS

$1 = \mathbf{1}$

$1 + 2 = \mathbf{3}$

$1 + 2 + 3 = \mathbf{6}$

$1 + 2 + 3 + 4 = \mathbf{10}$

$1 + 2 + 3 + 4 + 5 = \mathbf{15}$

$1 + 2 + 3 + 4 + 5 + 6 = \mathbf{21}$

$1 + 2 + 3 + 4 + 5 + 6 + 7 = \mathbf{28}$

$1 + 2 + 3 + 4 + 5 + 6 + 7 + 8 = \mathbf{36}$

$1 + 2 + 3 + 4 + 5 + 6 + 7 + 8 + 9 = \mathbf{45}$

$1 + 2 + 3 + 4 + 5 + 6 + 7 + 8 + 9 + 10 = \mathbf{55}$

The triangular numbers are the sums of successive counting numbers.

B

TRIANGULAR NUMBERS

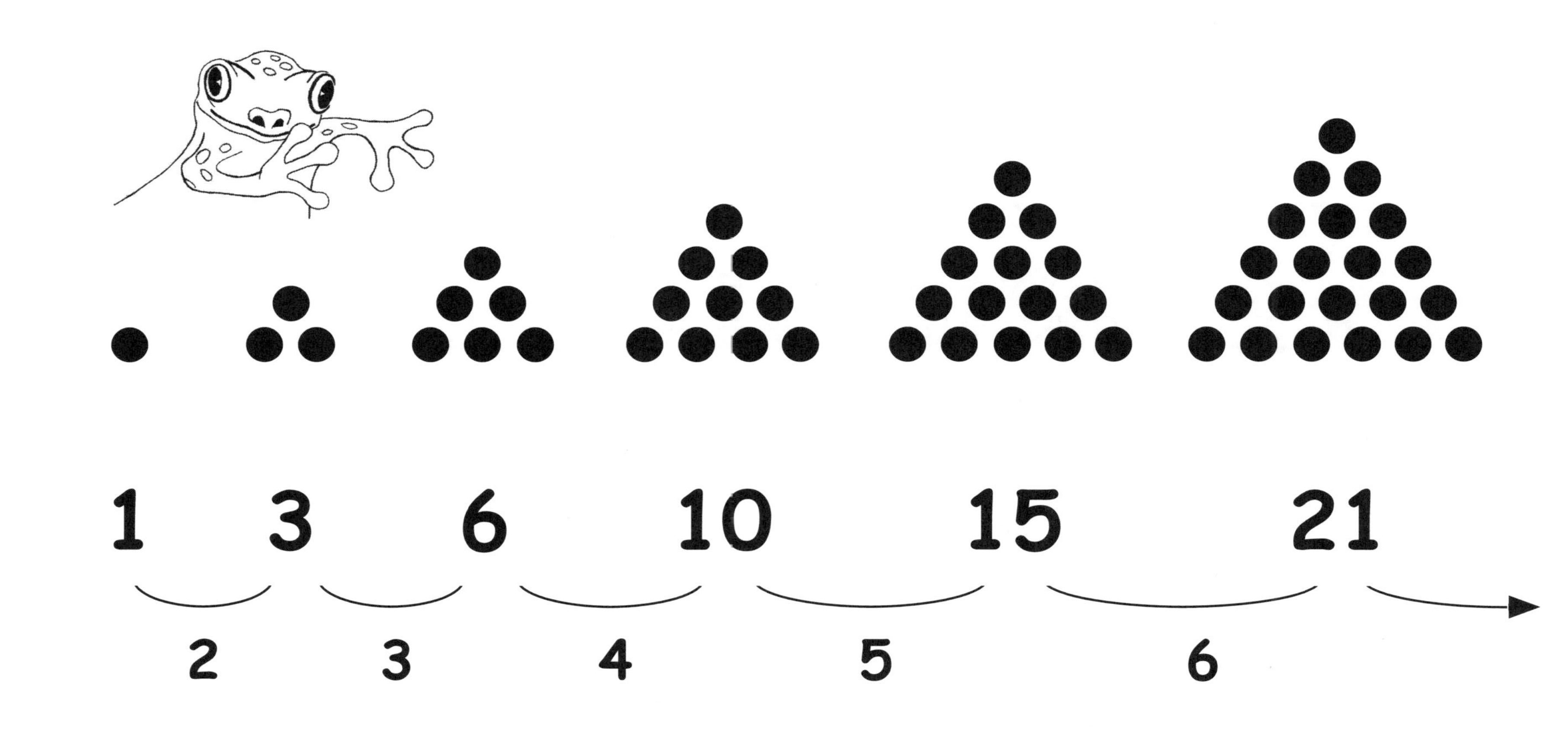

The triangular numbers can be represented as triangular arrays.

C

Eight paths that spell the word FROG

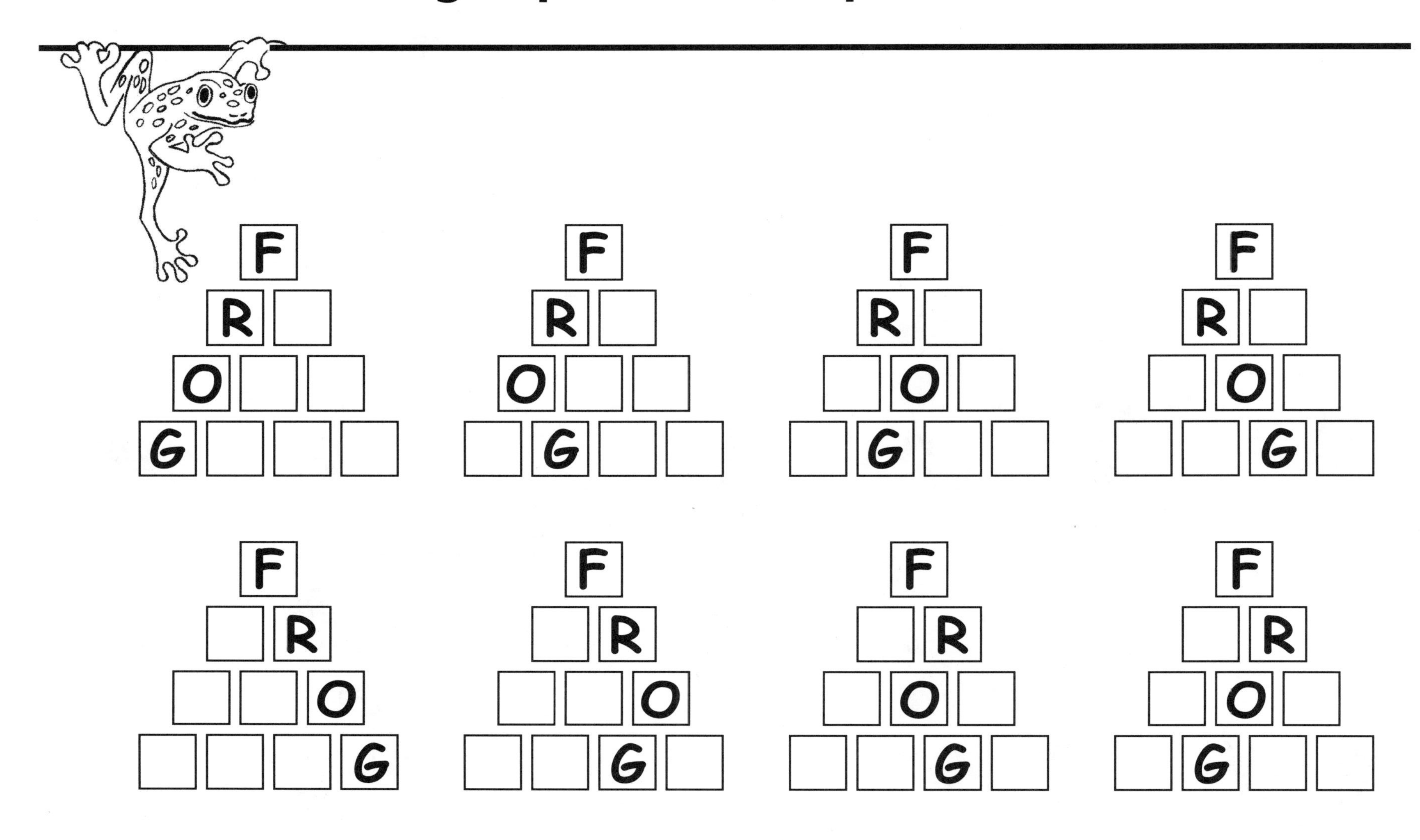

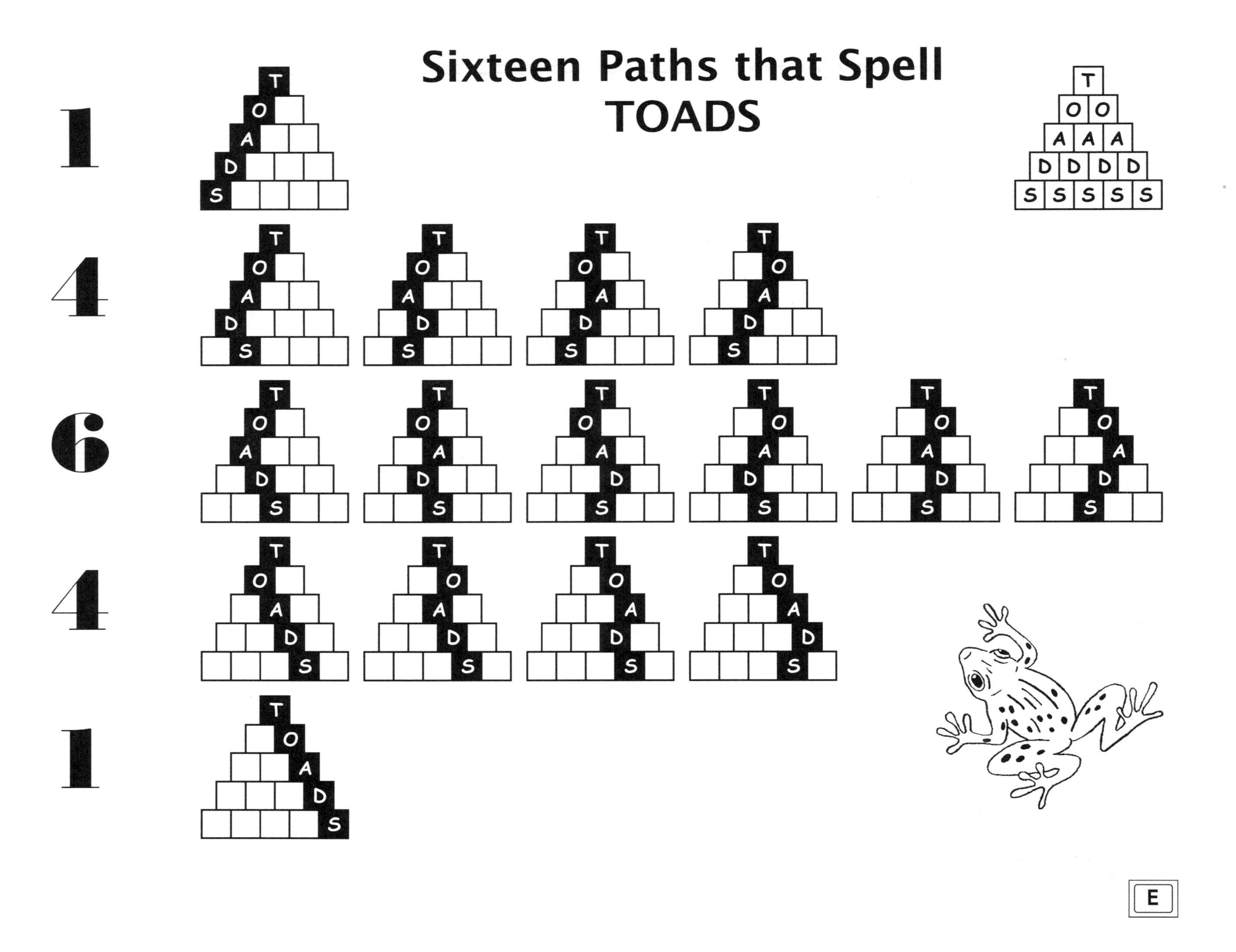

Sixteen Paths that Spell
TOADS
1
4
6
4
1
E

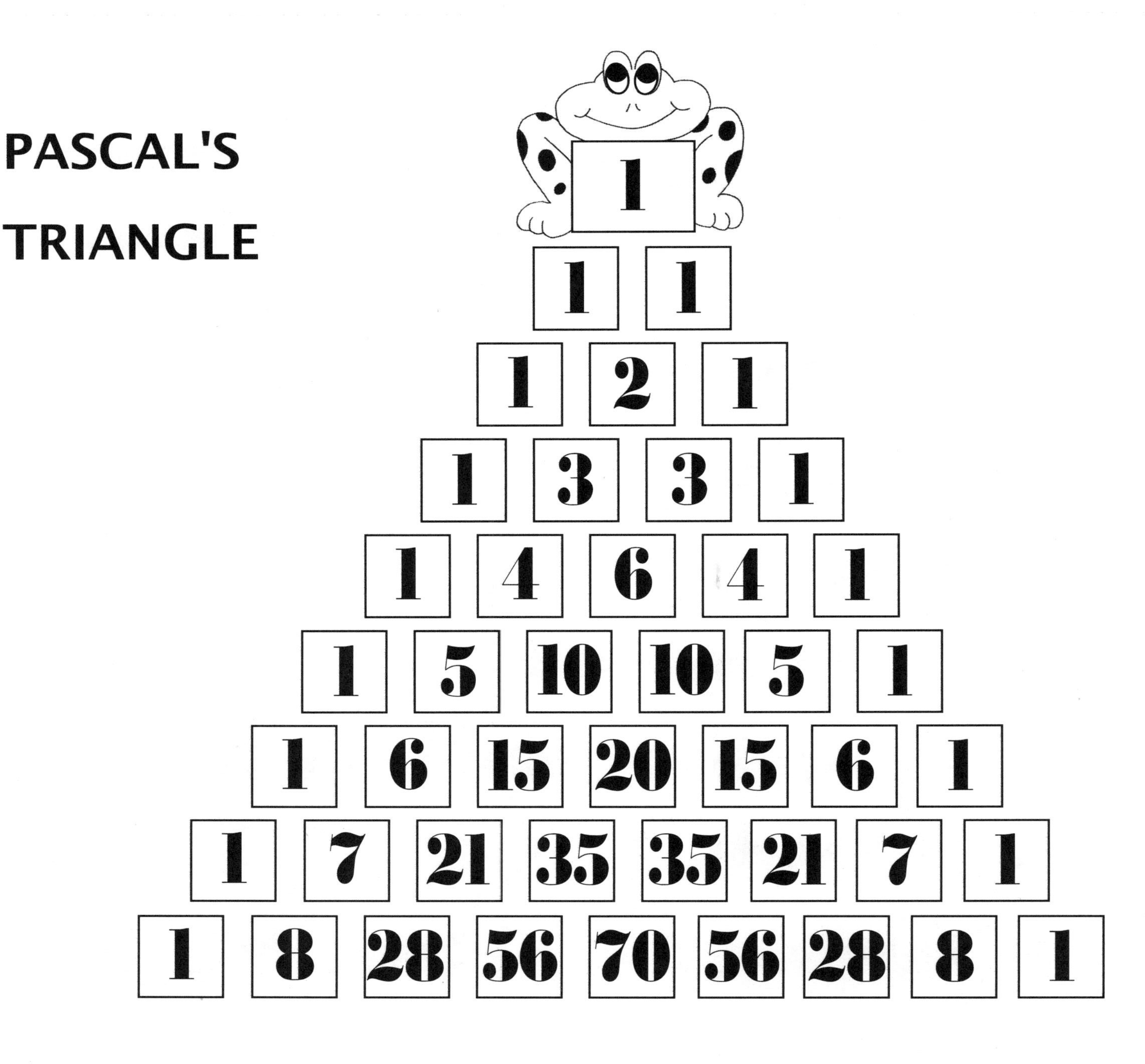
PASCAL'S
TRIANGLE
1
1 1
1 2 1
1 3 3 1
1 4 6 4 1
1 5 10 10 5 1
1 6 15 20 15 6 1
1 7 21 35 35 21 7 1
1 8 28 56 70 56 28 8 1

F

GETTING THE MOST from Activity 1

In the NCTM Standards, special attention is given to the importance of understanding the role that **communication** and **representation** play in the study of mathematics. Seeing and counting rectangles is one thing, but being able to identify and name them so that you can talk to others about them is quite another thing.

Have your students search for a number pattern as the paper strip is folded into different numbers of regions. Ask them to write down the process they use. What they put on paper can often reveal the level of their understanding of the generating procedure. The work of this student shows that he knows how to find successive **triangular numbers**.

1, 3, 6, 10, 15, 21, 28, 36
+2 +3 +4 +5 +6 +7 +8

The best way for students to identify a particular path down a triangular array of letters is to have them indicate, at each step, if the move down is to the left (L) or to the right (R). Encourage students to use this representation when describing a particular path.

Here are the three different paths that lead to the G in the second position from the left on the bottom. Each distinct path is represented by its own sequence of two L's and one R.

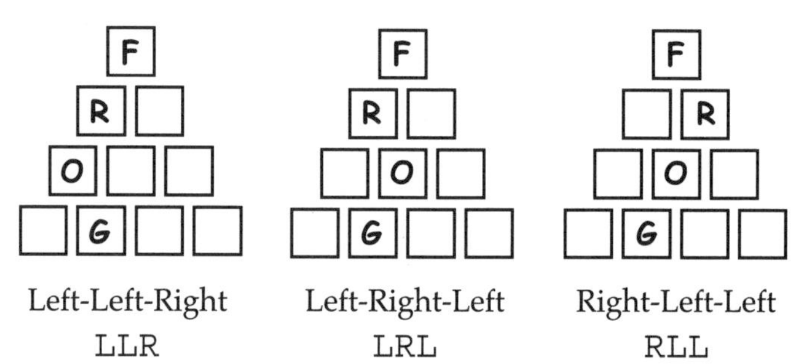

A clever way to randomly choose any one of the 8 possible paths down the array is to toss a coin 3 times. For heads, move left. For tails, move right.

HHH	HHT	HTH	THH	HTT	THT	TTH	TTT
LLL	LLR	LRL	RLL	LRR	RLR	RRL	RRR

Toss a coin 4 different times to randomly choose one of the 16 different paths down a triangular array for a 5-letter word.

STUDENT REACTIONS to Activity 1

After the folded-strip activity is completed, have your students write about their experience with **triangular numbers**. This not only gives them an opportunity to express their thoughts and reactions, but it also gives you a chance to assess your teaching.

I liked it because the pattern was cool.

Encourage students to enhance their visualization skills. It is important for them to be able to see things in their minds as well as with their eyes.

folding was hard because you couldn't see a thing

Much of mathematics centers around a good, strong understanding of concepts as well as proficiency with the basic skills. At the same time, we must foster **problem solving**. One method is through pattern recognition and extension.

Encourage your students to count the different paths down triangular arrays built from the letters in various words. Before they are through with their exploration, they will discover another important number pattern, *Pascal's triangle.*

```
   F
  R R
 O O O
G G G G
```

Paths to each G.
1 3 3 1

```
    T
   O O
  A A A
 D D D D
S S S S S
```

Paths to each S.
1 4 6 4 1

I liked Pascals triangle because it will never ever end, like numbers.

Each number in Pascal's triangle is the sum of the two numbers directly above it. This student saw quite a different pattern as well.

I thought I saw the square numbers and then I noticed that the triangular numbers were right there.

Triangular Numbers

```
                1
              1   1
            1   2   **1**
          1   3   **3**   1
        1   4   **6**   4   1
      1   5  **10**  10   5   1
    1   6  **15**  20  15   6   1
  1   7  **21**  35  35  21   7   1
1   8  **28**  56  70  56  28   8   1
.   .   .   .   .   .   .   .   .   .
```

Pascal's Triangle

SYMMETRY

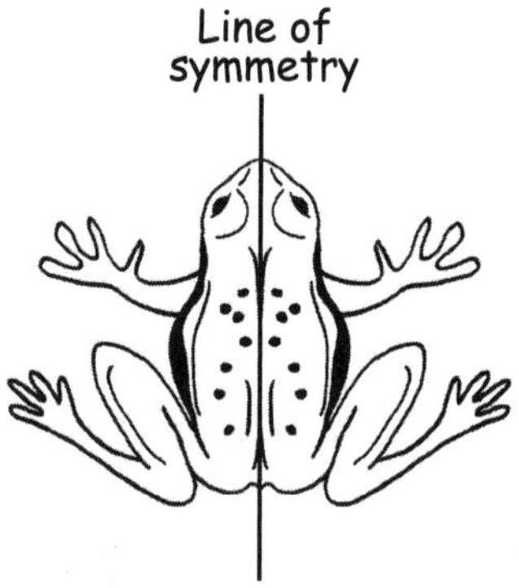

Line of symmetry

A frog's body is symmetric about a line.

PATTERNS in Letters

Introduction to Activity **2**

ARRANGING LETTERS
Rearranging Letters and Counting Choices

Description

This activity brings together all the NCTM process standards of problem solving, reasoning, connections, communication, and representation. Students work with physical objects that they can arrange and rearrange as they explore, apply, and reflect on a procedure for counting all possible orderings for any given set.

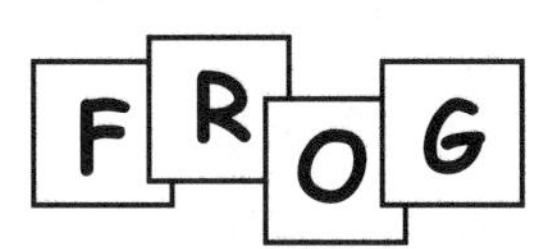

Six arrangements starting with F

FROG
FRGO
FORG
FOGR
FGRO
FGOR

Counting is one of the fundamental mathematical skills. While the activity begins with the creation of an organized list of possible arrangements, it ends with a generalized procedure for counting that makes use of the multiplication property of counting and the mathematical notation of factorials. Through this activity students gain a new view and a powerful application of numbers and the computational skill of multiplication.

Big Idea

Factorials are often used in counting different arrangements of distinct objects. Write n factorial as $n!$. It is the product of all successively decreasing counting numbers from n down through the number 1. For example, $5! = 5 \times 4 \times 3 \times 2 \times 1 = 120$.

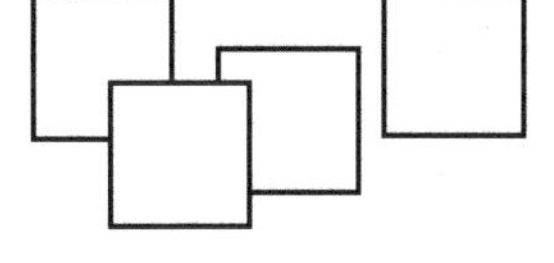

4 choices for the first letter,
3 for the second,
2 for the third,
and
1 for the fourth

$\mathbf{4! = 4 \times 3 \times 2 \times 1 = 24}$
possible arrangements in all

Expected Outcomes

Students will gain experience in

- creating organized lists
- counting systematically
- maintaining multiplication skills
- applying the multiplication property of counting
- using factorial notation

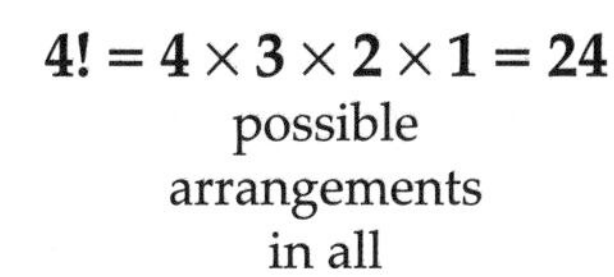

Meeting these NCTM Standards

✓ Numbers	✓ Problem Solving
Algebra	✓ Reasoning and Proof
Geometry	✓ Communication
Measurement	✓ Connections
✓ Data Analysis	✓ Representation

Activity 2 ARRANGING LETTERS

Getting Ready

Start with a 2 × 8-inch strip of paper.

Fold it in half and in half again.

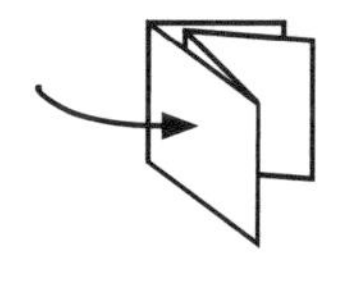

Unfold the strip.

Mark the four squares with the letters F, R, O, G.

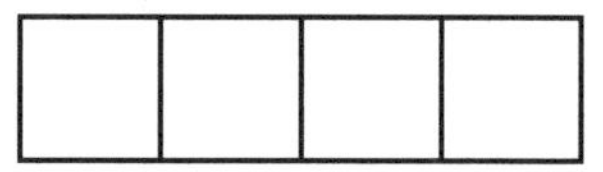

Cut the strip apart into four separate pieces.

Problem

One arrangement of these letters spells FROG.

Rearrange them in a row in some other ways.

How many different arrangements of these four letters are possible?

Remember, they do not have to spell a meaningful word. Students might choose to use an organized list as a **representation** of the different arrangements.

Solution

There are 24 possible arrangements.

For each of the 4 choices for the first letter, there are 3 choices for the second, 2 choices for the third, and 1 choice for the fourth. Multiply to find the total number.

FROG	FRGO	FORG	FOGR	FGRO	FGOR
RFOG	RFGO	ROFG	ROGF	RGFO	RGOF
OFRG	OFGR	ORFG	ORGF	OGFR	OGRF
GFRO	GFOR	GRFO	GROF	GOFR	GORF

$$4! = 4 \times 3 \times 2 \times 1 = 24$$

The product, 4 × 3 × 2 × 1, can be written in **factorial** form as 4!, read *four factorial*.

Note that if you were to arrange any 3 of the 4 letters in the word, FROG, there would still be 24 different possibilities. Just drop the last letter from each of the ones listed above.

FRO	FRG	FOR	FOG	FGR	FGO
RFO	RFG	ROF	ROG	RGF	RGO
OFR	OFG	ORF	ORG	OGF	OGR
GFR	GFO	GRF	GRO	GOF	GOR

$$4 \times 3 \times 2 = 24$$

Extensions

How many different ways can you arrange the 5 letters that are in the word TOADS?

There are 5 choices for the first letter, 4 for the second, 3 for the third, 2 for the fourth, and 1 for the fifth. Express the answer as 5 **factorial**. Multiply to find the total of 120.

$5! = 5 \times 4 \times 3 \times 2 \times 1 = 120$

The justification for multiplying is found in the *multiplication property for counting*.

If one event can occur in m ways and another in n ways, then the two events together can occur in $m \times n$ ways.

Both frogs and toads love to sit on lily pads.

How many different ways can you arrange the 4 letters that are in the word LILY? How many ways can you arrange the 3 letters in the word PAD? Do some careful **reasoning**.

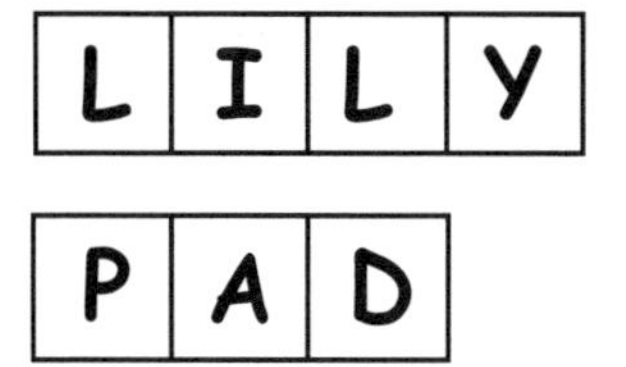

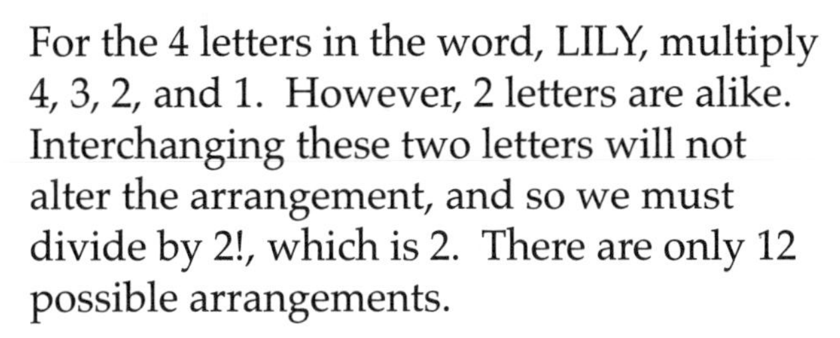

For the 4 letters in the word, LILY, multiply 4, 3, 2, and 1. However, 2 letters are alike. Interchanging these two letters will not alter the arrangement, and so we must divide by 2!, which is 2. There are only 12 possible arrangements.

$4! = 4 \times 3 \times 2 \times 1 = 24$

$24 \div 2 = 12$

LILY LIYL LLIY LLYI LYLI LYIL
ILLY ILYL IYLL YLIL YLLI YILL

For the 3 letters in the word, PAD, use 3!. There are just 6 possible arrangements.

$3! = 3 \times 2 \times 1 = 6$

PAD PDA APD ADP DPA DAP

Cooperative Activity

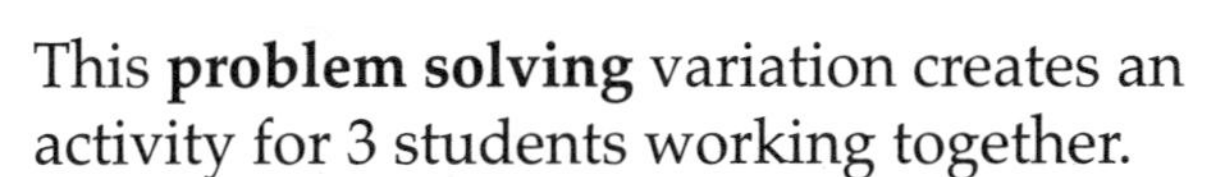

This **problem solving** variation creates an activity for 3 students working together.

If any letter in the word PAD can be repeated any number of times, how many 3-letter arrangements are possible?

$3^3 = 3 \times 3 \times 3 = 27$

PPP AAA DDD

PPA PAP APP PPD PDP DPP
AAP APA PAA AAD ADA DAA
DDP DPD PDD DDA DAD ADD

PAD PDA APD ADP DPA DAP

ARRANGING LETTERS **Worksheet 2A**

Start with a 2 x 8-inch strip of paper.
Fold it in half and in half again. Then unfold the strip.
Mark the four squares with the letters F, R, O, and G.

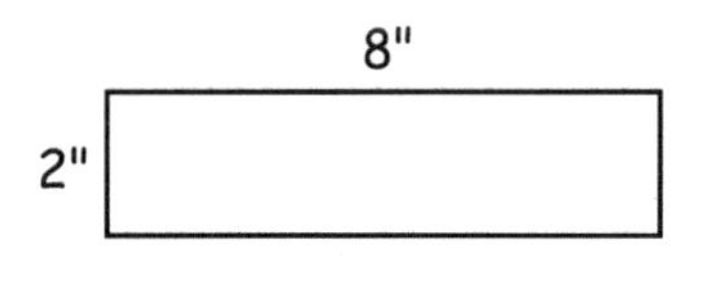

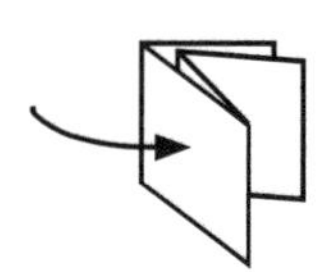

Now cut the strip into four separate lettered pieces.
Three arrangements of these four letters are shown below.

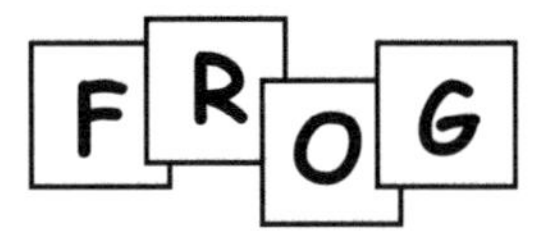

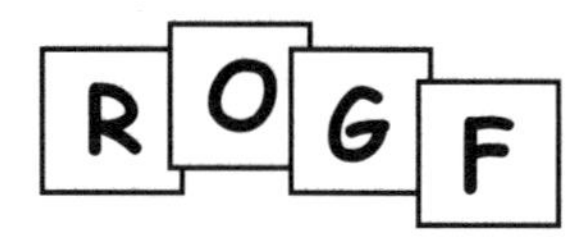

1. Find all the arrangements possible that start with F, R, O, or G.

F _ _ _ **R** _ _ _ **O** _ _ _ **G** _ _ _

2. In all, how many different arrangements can you find for the four letters in the word FROG? ______

3. How many arrangements of the five letters in the word FROGS start with the letter S? Think carefully. ______

4. In all, how many arrangements do you think are possible using all five letters in the word FROGS? ______

FACTORIALS

Worksheet 2B

.The product, 4 x 3 x 2 x 1, can be written as 4!, read **four factorial**.

1. Three factorial means 3 x 2 x 1. Find the value of 3!. _____

2. Five factorial means 5 x 4 x 3 x 2 x 1. Find the value of 5!. _____

3. Find the value of 6!. _____

Start with a word, any word, with four different letters. There will always be 4! or 24 different ways that those four letters can be arranged in a row. The reasoning goes like this:

> For each of the 4 choices for the first letter, there are 3 choices for the second, 2 choices for the third, and 1 choice for the fourth. Multiply to find the total number of choices.

4. The word TOAD has four different letters. They can be arranged in 4! different ways. How many ways is that? _____

5. Use a factorial to express the total number of different ways the five letters in the word TOADS can be arranged. Then find its value. _____ _____

6. Find the number of ways the letters in each word can be arranged. Express each answer as a factorial, and then find its value.

   LEGS _____ _____

   TONGUE _____ _____

   STOMACH _____ _____

7. There are only 12 ways to arrange the four letters in the word EYES. List the 12 arrangements. Can you explain why there are not 24 choices as there are with the word FROG?

ANSWERS to Worksheets 2A and 2B

ARRANGING LETTERS — Worksheet 2A

Start with a 2 x 8-inch strip of paper.
Fold it in half and in half again. Then unfold the strip.
Mark the four squares with the letters F, R, O, and G.

8"
2"

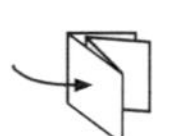

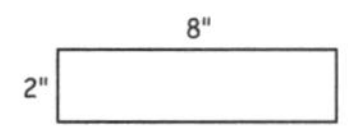

Now cut the strip into four separate lettered pieces.
Three arrangements of these four letters are shown below.

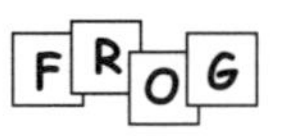

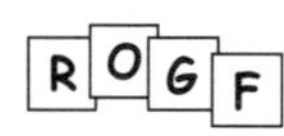

1. Find all the arrangements possible that start with F, R, O, or G.

F _ _ _	R _ _ _	O _ _ _	G _ _ _
FROG	RFOG	OFRG	GFRO
FRGO	RFGO	OFGR	GFOR
FORG	ROFG	ORFG	GRFO
FOGR	ROGF	ORGF	GROF
FGRO	RGFO	OGFR	GOFR
FGOR	RGOF	OGRF	GORF

2. In all, how many different arrangements can you find for the four letters in the word FROG? **24**

3. How many arrangements of the five letters in the word FROGS start with the letter S? Think carefully. **24**

4. In all, how many arrangements do you think are possible using all five letters in the word FROGS? **120**

20

FACTORIALS — Worksheet 2B

The product, 4 x 3 x 2 x 1, can be written as 4!, read **four factorial**.

1. Three factorial means 3 x 2 x 1. Find the value of 3!. **6**

2. Five factorial means 5 x 4 x 3 x 2 x 1. Find the value of 5!. **120**

3. Find the value of 6!. **720**

Start with a word, any word, with four different letters. There will always be 4! or 24 different ways that those four letters can be arranged in a row. The reasoning goes like this:

For each of the 4 choices for the first letter, there are 3 choices for the second, 2 choices for the third, and 1 choice for the fourth. Multiply to find the total number of choices.

4. The word TOAD has four different letters. They can be arranged in 4! different ways. How many ways is that? **24**

5. Use a factorial to express the total number of different ways the five letters in the word TOADS can be arranged. Then find its value. **5!** **120**

6. Find the number of ways the letters in each word can be arranged. Express each answer as a factorial, and then find its value.

LEGS	**4!**	**24**
TONGUE	**6!**	**720**
STOMACH	**7!**	**5040**

7. There are only 12 ways to arrange the four letters in the word EYES. List the 12 arrangements. Can you explain why there are not 24 choices as there are for the word FROG? **Two letters are the same.**

EYES	EEYS	EYSE	YEES	YESE	YSEE
ESEY	EESY	ESYS	SEEY	SEYE	SYEE

21

FROG Facts

There are about 4000 different species of frogs. You can find frogs virtually everywhere in the United States, but the widest assortment of different kinds of frogs can be found in Africa. The only continent where frogs cannot be found is Antarctica.

TRANSPARENCY MASTERS

Make transparencies from these masters.

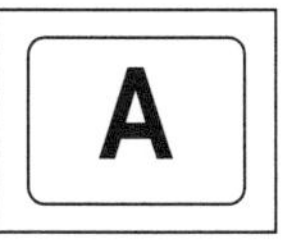

Make two copies of this transparency. Keep one copy and cut out the separate letters from the other. Encourage your students to come up and move these letters around on the blank grid to illustrate different arrangements of letters.

Use this transparency to show the 24 different ways that the 4 different letters in the word FROG can be arranged. Since there are 4 choices for the first letter, 3 for the second, 2 for the third, and 1 for the fourth, the total number, 24, can be found by multiplying these four numbers. Another way to express this product is 4!.

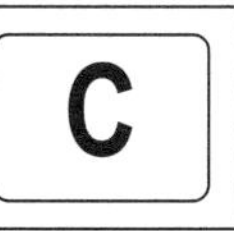

This transparency shows the 24 different ways any 3 of the 4 different letters in the word FROG can be arranged. Note that this list is the same as that on transparency A except that, in each case, the fourth letter has not been shown.

To assess whether students understand this process, ask at the end of the discussion for the number of different ways any 2 of these 4 letters can be arranged. In this case, the answer is $4 \times 3 = 12$.

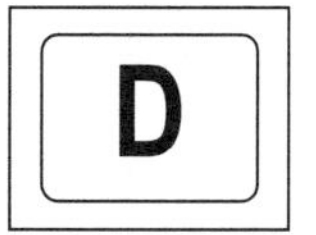

Use this transparency when discussing factorial notation. Students need to recognize that factorials increase in value very rapidly. 10! is already well over three and a half million. This means that 10 different letters can be arranged in a row in 10! or 3,628,800 different ways. Writing a new arrangement every second, day and night, would take a full 42 days!

ARRANGING LETTERS

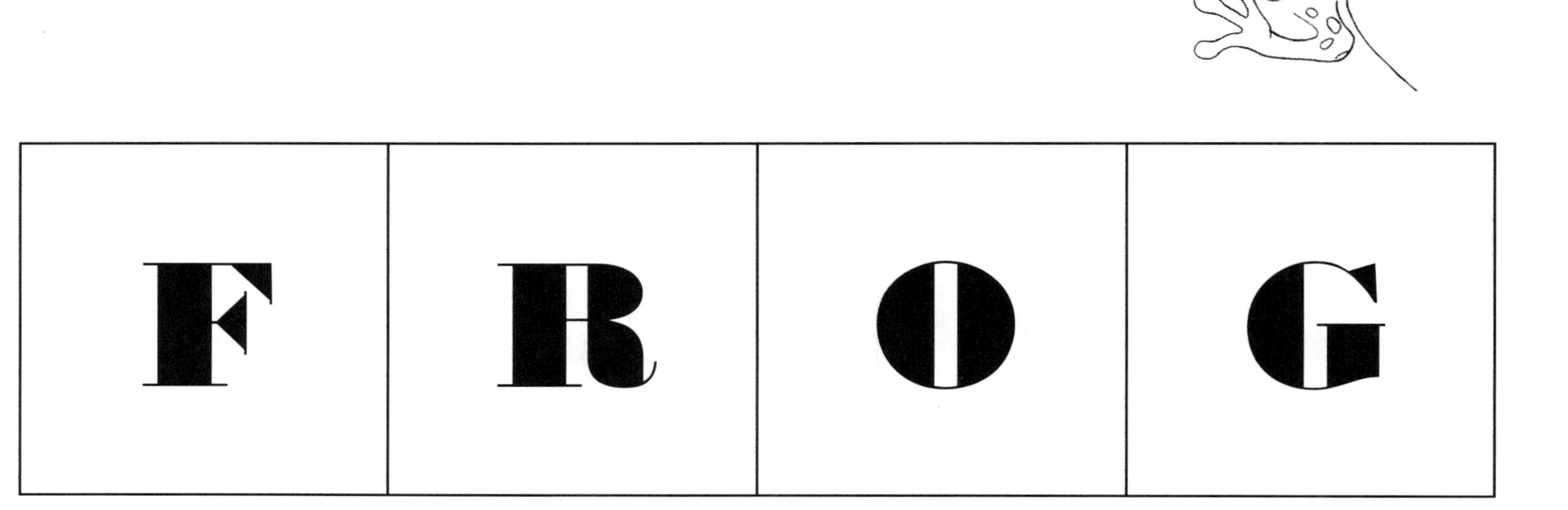

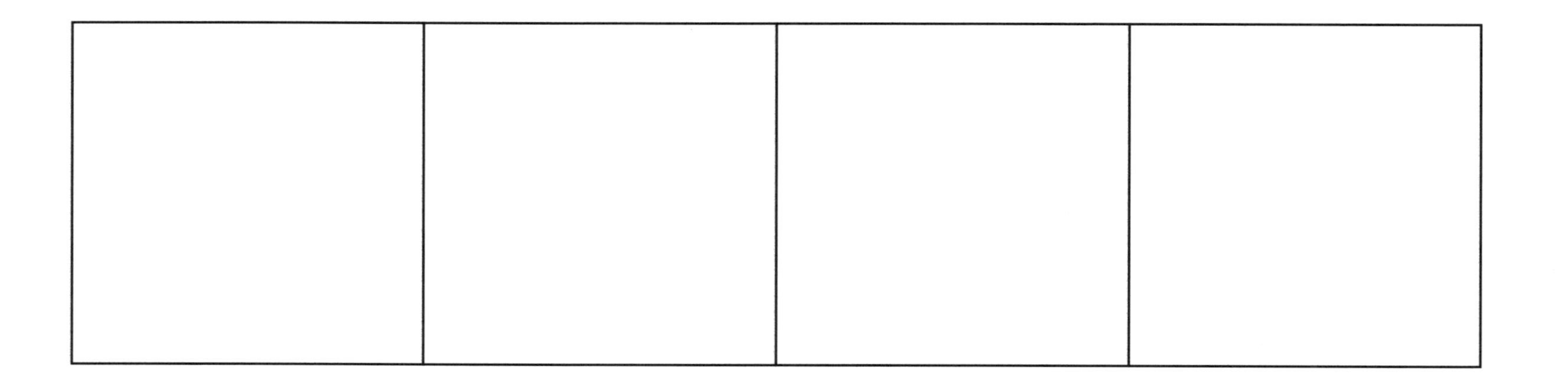

A

ARRANGING THE 4 LETTERS IN FROG

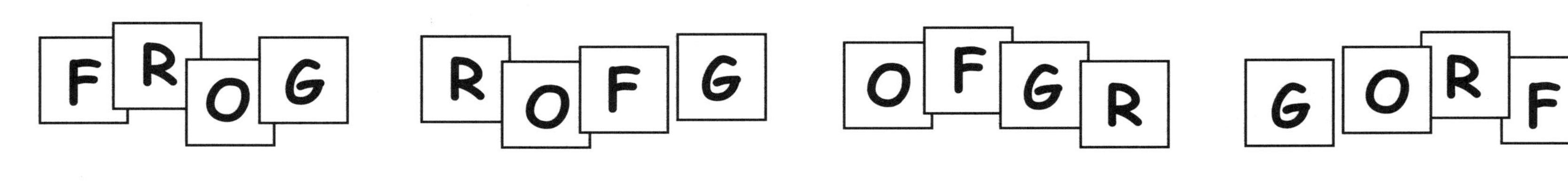

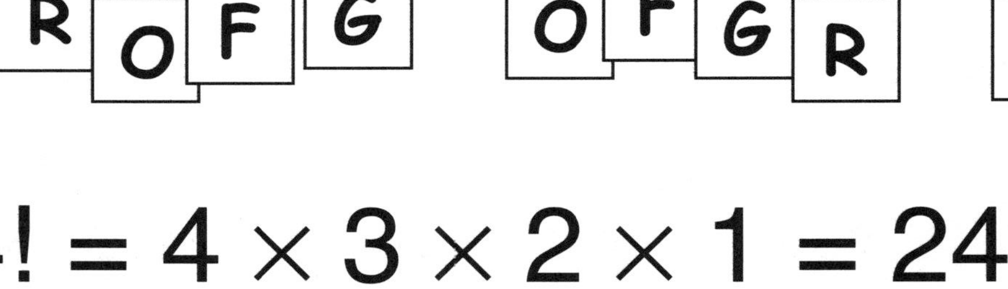

$$4! = 4 \times 3 \times 2 \times 1 = 24$$

FROG	FRGO	FORG	FOGR	FGRO	FGOR
RFOG	RFGO	ROFG	ROGF	RGFO	RGOF
OFRG	OFGR	ORFG	ORGF	OGFR	OGRF
GFRO	GFOR	GRFO	GROF	GOFR	GORF

4 different letters can be arranged
in 24 different ways.

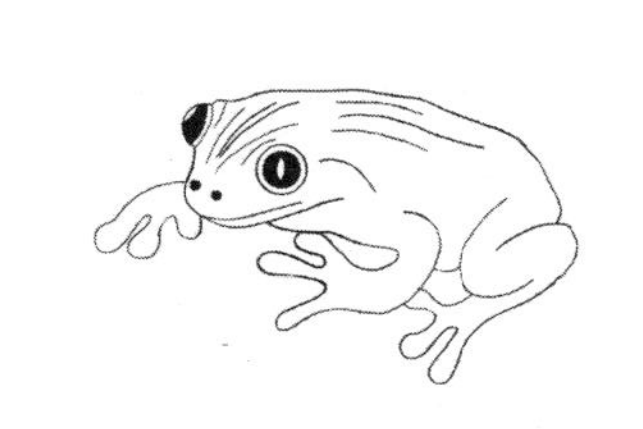

B

ARRANGING ANY 3 LETTERS IN FROG

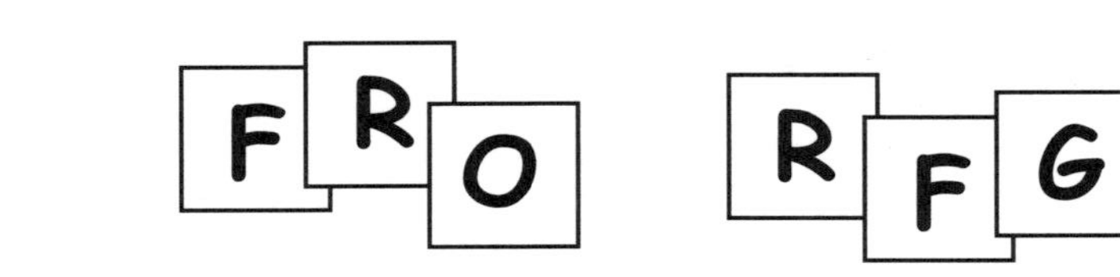
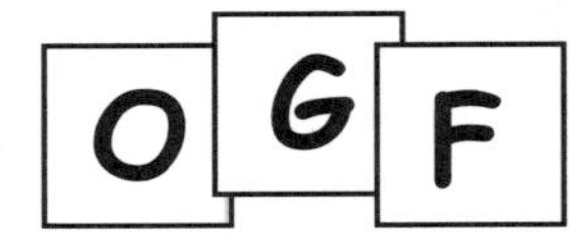

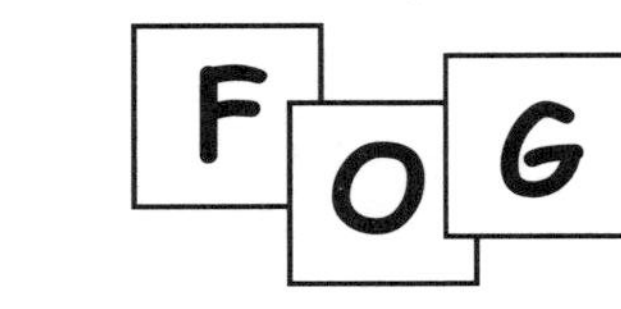

$$4 \times 3 \times 2 = 24$$

FRO	FRG	FOR	FOG	FGR	FGO
RFO	RFG	ROF	ROG	RGF	RGO
OFR	OFG	ORF	ORG	OGF	OGR
GFR	GFO	GRF	GRO	GOF	GOR

4 different letters can be arranged,
3 at a time, in 24 different ways.

C

FACTORIALS

$$1! = 1 = 1$$

$$2! = 2 \times 1 = 2$$

$$3! = 3 \times 2 \times 1 = 6$$

$$4! = 4 \times 3 \times 2 \times 1 = 24$$

$$5! = 5 \times 4 \times 3 \times 2 \times 1 = 120$$

$$6! = 6 \times 5 \times 4 \times 3 \times 2 \times 1 = 720$$

$$7! = 7 \times 6 \times 5 \times 4 \times 3 \times 2 \times 1 = 5{,}040$$

$$8! = 40{,}320$$

$$9! = 362{,}880$$

$$10! = 3{,}628{,}800$$

D

GETTING THE MOST from Activity 2

One way to begin this activity is to let students try to find as many arrangements as possible by simply manipulating the four lettered squares.

I physically moved the squares.

Any 4 letters can be arranged in 24 different ways as long as they are different letters. There are 4 choices for the first letter, 3 for the second, 2 for the third, and 1 left for the fourth. The product of these numbers, called 4 **factorial**, is 24.

$$4! = 4 \cdot 3 \cdot 2 \cdot 1 = 24$$

Factorials can be used to express the total number of possible arrangements of any number of letters, as long as the letters are different and all of them are used.

Here is a sample of a student's work in analyzing data where only 3 letters of the word FROG are to be used in quite a different **problem-solving** situation.

The probability that a randomly chosen arrangement of any 3 of the 4 letters in the word FROG spell out a meaningful word is 1/8.

The probability that a random arrangement of any 2 of the 4 letters in the word FROG spell out a meaningful word is 3/12 or 1/4, as shown below.

(FOG)	FGR	(FOR)	GOR
FGO	FRG	(FRO)	GRO
GFO	RGF	ROF	ROG
GOF	RFG	RFO	RGO
OFG	GFR	OFR	~~(ORG)~~ ORG
OGF	GRF	ORF	OGR

probability = 3/24 = 1/8 chance

FO FG FR RF RG RO (OF) OG (OR) GF (GO) GR

For an interesting **connection** to alphabetizing, see the Problem Solving Puzzle A on page 110. In this puzzle students are asked to both list and alphabetize various arrangements of letters from the word FROG.

STUDENT REACTIONS to Activity 2

Strengthen **communication** and **reasoning** skills in your students by encouraging them to express their thoughts through writing. These comments illustrate some of the ways students tackled the problem of listing the 24 arrangements for the four different letters in FROG.

The way I got my list was that I took the word FROG and then I shuffled the letters around to form new words.

I randomly moved letters around.

$6 \times 4 = 24$

I split them into four sets. the sets that begin with F, R, O, and G. Each set has 6 different orders.

Students need to be able to make systematic, organized lists. This is an important **problem-solving** skill. The list of the 24 arrangements, shown below, is correct, but no clear systematic process of listing beyond the first letter is apparent.

Add a fifth letter, different from the first four, and see if your students recognize that they will have five times as many possible arrangements.

F	R	O	G
FROG	ROFG	ORFG	GRFO
FORG	RFOG	OGFR	GFRO
FRGO	RGFO	OGRF	GROF
FOGR	RFGO	OFRG	GORF
FGRO	RGOF	OFGR	GOFR
FGOR	ROGF	ORGF	GFOR

Frogs $5! = 5 \times 4 \times 3 \times 2 \times 1 = 120$

If the 5-letter arrangements can use any of the 5 letters any number of times, then the multiplication property of counting give quite a different answer.

$$5^5 = 5 \times 5 \times 5 \times 5 \times 5 = 3{,}125$$

MATH and FROGS

MULTIPLICATION

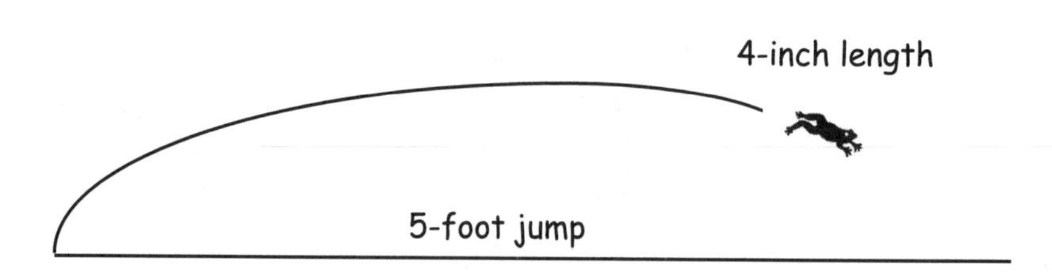

The leopard frogs of North America are great jumpers. A 4-inch leopard frog can jump about 15 times its body length. That's 5 feet!

PATTERNS in Moves

Introduction to Activity 

PLANNING HOPS AND JUMPS
Playing by the Rules

Description

This activity on hops and jumps offers the students an opportunity to experience all five of the NCTM process standards of problem solving, reasoning, communication, connections, and representation. Students work with physical objects, moving them right and left in a grid according to a prescribed set of rules. They explore, search, discover, count, record, and then reflect on the specific procedures used through a variety of different representations.

Students play the game according to a specific set of rules. As they gain proficiency with the game, they need to replicate and record their moves, showing their solutions. This provides a valuable opportunity for discovery and discussion about different patterns and their extensions.

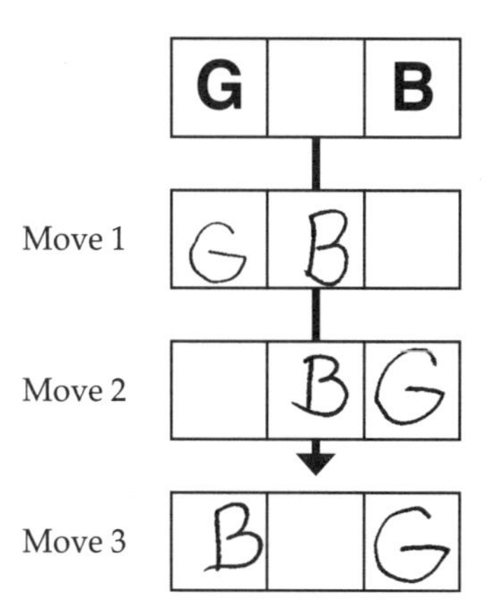

Only 3 hops and jumps are needed to reverse the positions.

Big Idea

For students to succeed in this game, they need to develop a good **thinking strategy**. This often begins with some trial and error moves that lead, through some analysis and logical thinking, to a final successful game plan.

Expected Outcomes

Students will gain experience in

- following specific directions
- searching for and extending patterns
- thinking and applying strategies
- recording results
- using different representations

Meeting these NCTM Standards

✓ Numbers	✓ Problem Solving
Algebra	✓ Reasoning and Proof
Geometry	✓ Communication
Measurement	✓ Connections
✓ Data Analysis	✓ Representation

Activity 3 PLANNING HOPS AND JUMPS

The Game

Call the game ***Moving Day at Lily Pad Pond.***

Four frogs are sitting on lily pads in a pond.

> Two brown frogs are on pads on the right.
> Two green frogs are on pads on the left.
> One pad separates them.

2 green frogs 2 brown frogs

G G B B

The frogs move by hops and by jumps.

The Rules

1. Only one frog can move at a time.
   Brown frogs on the right can only move to the left.
   Green frogs on the left can only move to the right.
2. Only two kinds of moves are allowed.
   Hop onto an adjacent empty pad.
   Jump over a single frog of the opposite color onto an empty pad.

The Record

Choose one of these forms of **representation** to use in recording your moves.

- Use lettered squares.
- Use sentences.
- Use the letters G and B.
  Circle those that are jumps.
- Use 1 for a hop and 2 for a jump.
  Designate if G or B moves first.

Green hops right.

Brown jumps left.

Brown hops left.

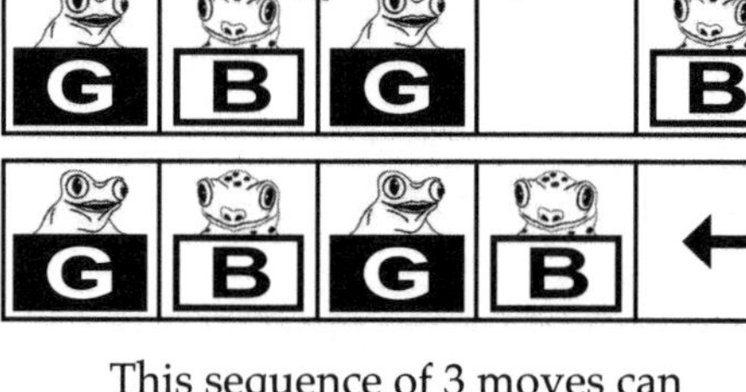

This sequence of 3 moves can also be represented as

G (B) B or 121.

The Problem

> *Follow the rules.*
> *Move the frogs to the opposite sides of the pond.*
> *Green frogs move to the right and brown frogs to the left.*
> *Try for the least number of moves.*

Bring your best **problem-solving** skills into play.
Remember, frogs must stay on the pads.

The Solution

A minimum of 8 moves are required. The solution is given here in words and in pictures, assuming green moves first.

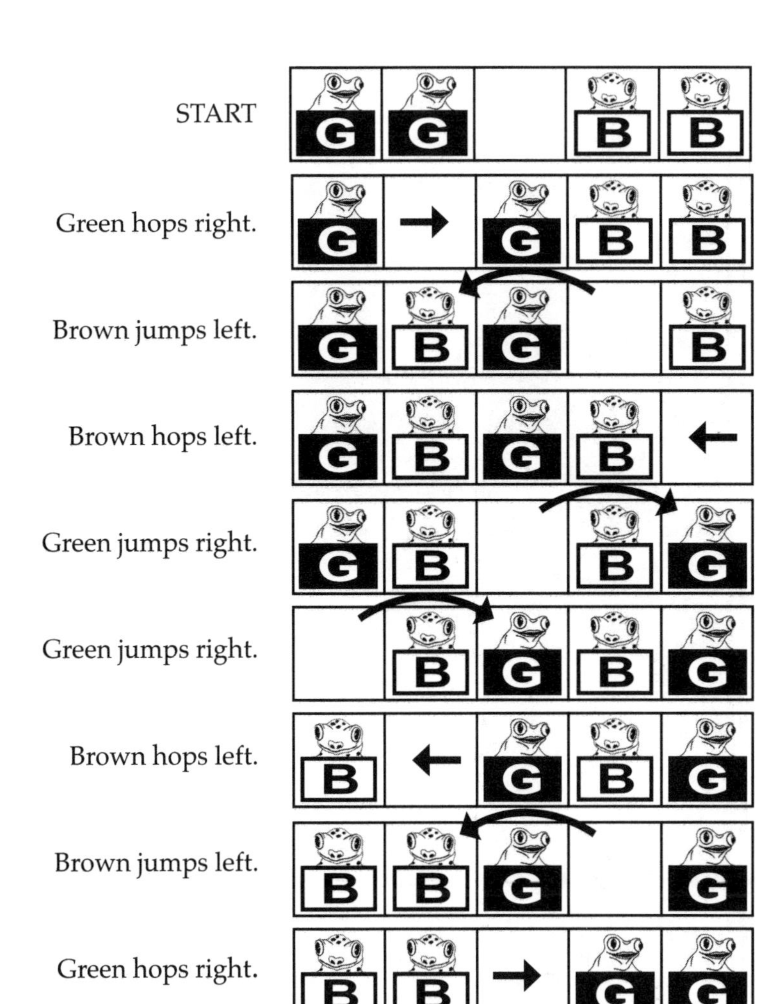

The same 8-step solution sequence can also be shown this way using only the letters G and B. Jumps are circled. Hops are not.

G (B) B (G)(G) B (B) G

Challenge your students to use **reasoning** alone to establish this 8-step solution when brown moves first.

B (G) G (B)(B) G (G) B

Another **representation** for both these solutions uses 1's for hops and 2's for jumps.

1 2 1 2 2 1 2 1

Extension

Have students apply their **thinking strategies** to other cases with different numbers of frogs. See if they can verify the entries in this table and discover some interesting patterns to help them predict the results for more frogs.

Frogs per side	Sequence of moves	Total number of moves
1	G (B) G	3
2	G (B) B (G)(G) B (B) G	8
3	G (B) B (G)(G) G (B)(B)(B) G (G)(G) B (B) G	15

For n frogs per side, the minimum total number of moves required is $(n + 1)^2 - 1$.

MOVING DAY
at Lily Pad Pond

Worksheet 3

- Only one frog can move at a time.
  Brown frogs move left and green frogs move right.
- Only two kinds of moves are allowed.
  Hop to an adjacent pad or jump over a frog of the opposite color.

There are five lily pads in a row in Lily Pad Pond. Two green frogs are on the pads on the left, and two brown frogs are on the pads on the right.

One at a time, they start hopping and jumping. In just 8 moves, the frogs reverse their positions.

Can you do it?

1. Move the two green frogs to the right and the two brown frogs to the left in just 8 moves. Let green move first, and follow the rules. Record each move in the table.

2. Write the same 8 successive moves as a sequence of 1's and 2's. Use a 1 for each hop in either direction and a 2 for each jump.

Start	G	G		B	B
Move 1					
Move 2					
Move 3					
Move 4					
Move 5					
Move 6					
Move 7					
Move 8					
End	B	B		G	G

3. Give a single sequence of the letters G and B to describe the same 8-step solution when a green frog moves first. Use G for a move of a green frog. Use B for a move of a brown frog. Circle the letter if it is a jump. Do not circle the letter if it is a hop.

____ ____ ____ ____ ____ ____ ____ ____

4. This time use the letters G and B to describe the 8-step solution when a brown frog moves first.

____ ____ ____ ____ ____ ____ ____ ____

Now consider a version of the game where 3 frogs start on each side. The rules for playing remain the same.

G	G	G		B	B	B

Here is the solution that shows the fewest moves necessary to reverse the positions of the frogs. Remember, a 1 means a hop, and a 2 means a jump.

1 2 1 2 2 1 2 2 2 1 2 2 1 2 1

5. How many moves are represented? ____

   How many of them are jumps? ____

6. Represent this solution as a sequence of the letters G and B. Circle the moves that are jumps. Assume green moves first.

7. Act out the solution given and verify that it does indeed reverse the positions of the two sets of 3 frogs.

ANSWERS to Worksheet 3

MOVING DAY
at Lily Pad Pond **Worksheet 3**

- Only one frog can move at a time.
  Brown frogs move left and green frogs move right.
- Only two kinds of moves are allowed.
  Hop to an adjacent pad or jump over a frog of the opposite color.

There are five lily pads in a row in Lily Pad Pond. Two green frogs are on the pads on the left, and two brown frogs are on the pads on the right.

One at a time, they start hopping and jumping. In just 8 moves, the frogs reverse their positions.

Can you do it?

1. Move the two green frogs to the right and the two brown frogs to the left in just 8 moves. Let green move first, and follow the rules. Record each move in the table.

2. Write successive moves as a sequence of 1's and 2's. Use a 1 for each hop in either direction and a 2 for each jump.

12122121

Start	G	G		B	B
Move 1	G		G	B	B
Move 2	G	B	G		B
Move 3	G	B	G	B	
Move 4	G	B		B	G
Move 5		B	G	B	G
Move 6	B		G	B	G
Move 7	B	B	G		G
Move 8	B	B		G	G
End	B	B		G	G

34

3. Give a single sequence of the letters G and B to describe the same 8-step solution when a green frog moves first. Use G for a move of a green frog. Use B for a move of a brown frog. Circle the letter if it is a jump. Do not circle the letter if it is a hop.

G (B) B (G) (G) B (B) G

4. This time use the letters G and B to describe the 8-step solution when a brown frog moves first.

B (G) G (B) (B) G (G) B

Now consider a version of the game where 3 frogs start on each side. The rules for playing remain the same.

G	G	G		B	B	B

Here is the solution that shows the fewest moves necessary to reverse the positions of the frogs. Remember, a 1 means a hop, and a 2 means a jump.

1 2 1 2 2 1 2 2 2 1 2 2 1 2 1

5. How many moves are represented? **15**

   How many of them are jumps? **9**

6. Represent this solution as a sequence of the letters G and B. Circle the moves that are jumps. Assume green moves first.

G (B) B (G) (G) G (B) (B) (B) G (G) (G) B (B) G

7. Act out the solution given and verify that it does indeed reverse the positions of the two sets of 3 frogs.

35

FROG Facts

Perhaps the most famous frog-jumping contest is held in California. It draws some 2000 contestants annually, and is an outgrowth of Mark Twain's famous tale, ***The Celebrated Jumping Frog of Calaveras County***. The strong hind legs of some frogs enable them to jump as much as 40 times their length.

TRANSPARENCY MASTERS

Make transparencies from these masters.

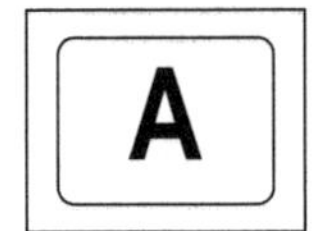

Make two copies of this transparency. Use the first copy as is. Cut four separate squares from the second copy, each one containing a single frog, two green and two brown.

Use several of the cut squares to demonstrate on the grid the difference between a hop and a jump. Remind students of the two different directions that the two kinds of frogs must move.

After the students have played the game on the first page of worksheet 4 themselves, let individuals demonstrate their solutions using the transparency. Alternatively, you may choose to illustrate yourself the 8-step solution using the least number of moves.

This transparency shows the complete set of 8 moves where green moves first. On the left of the transparency, the moves are described verbally. On the right, successive moves are identified by the letters G and B, adding a new letter to the sequence at each step. Circled letters indicate moves that are jumps instead of hops. Valuable experience is gained by showing students the solution using different representations.

MOVING DAY AT LILY PAD POND

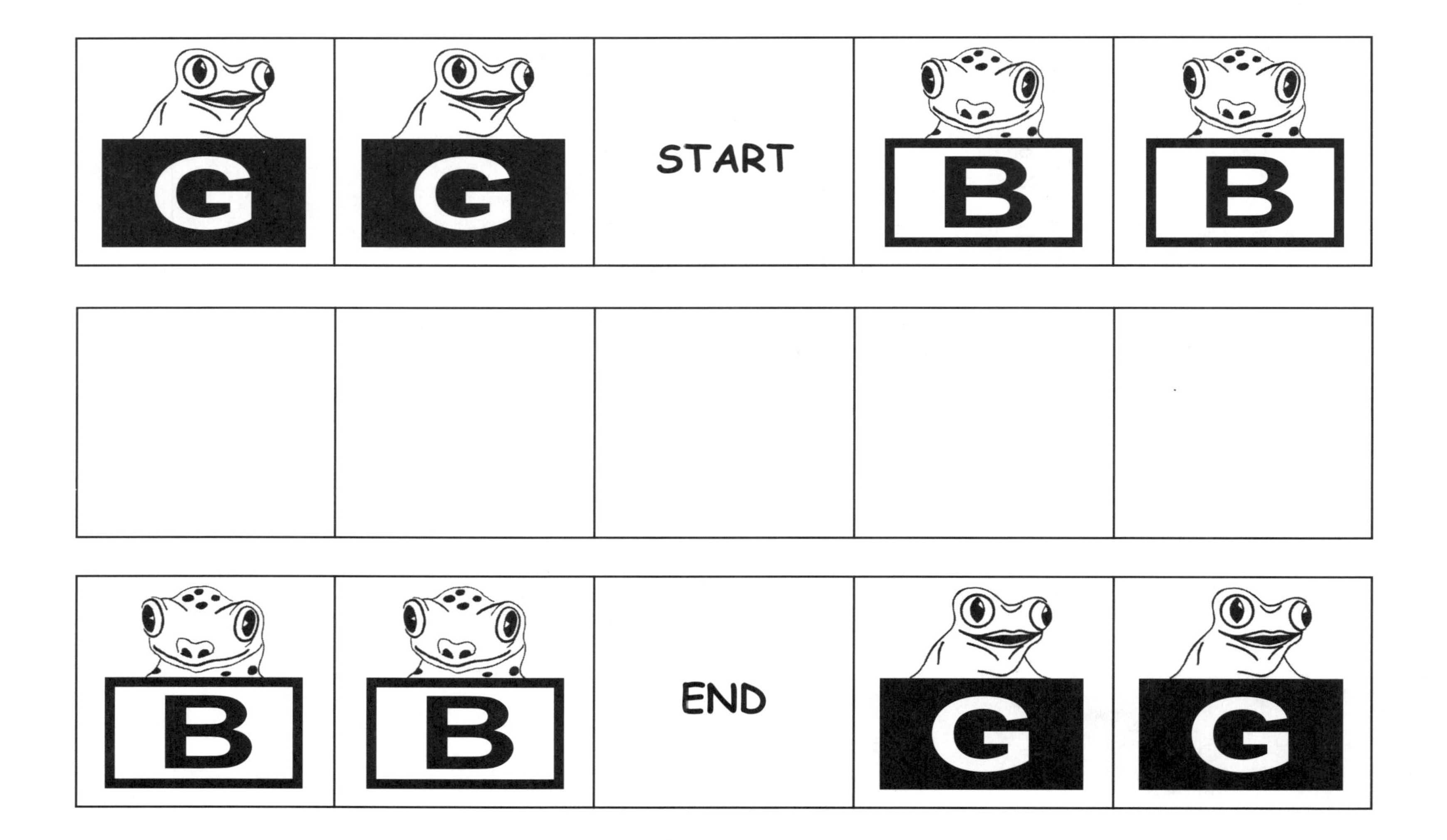

A

MOVING DAY at Lily Pad Pond

Move	1	2	3	4	5	Record
Start.	G	G		B	B	
Green hops right.	G	→	G	B	B	G
Brown jumps left.	G	B	G		B	G (B)
Brown hops left.	G	B	G	B	←	G (B) B
Green jumps right.	G	B		B	G	G (B) B (G)
Green jumps right.		B	G	B	G	G (B) B (G)(G)
Brown hops left.	B	←	G	B	G	G (B) B (G)(G) B
Brown jumps left.	B	B	G		G	G (B) B (G)(G) B (B)
Green hops right.	B	B	→	G	G	G (B) B (G)(G) B (B) G

B

GETTING THE MOST from Activity 3

Some students may not understand how to record their moves in the table on the first page of the worksheet. The process may be worth practicing first with just one frog on each side. However, for other students, starting with just one frog of each color may take away from the challenge of the activity.

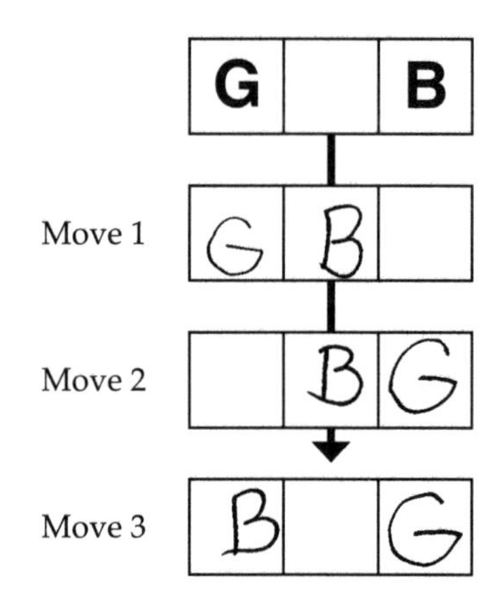

If you start with one frog of each color it will give ~~the~~ trick away.

In the solution above, green can only move right and brown can only move left. For another **representation** of the solution, use R for a right and L for a left move.

Here are the 3 moves for two frogs, one on each side.

green first	**RLR**	1 right, 1 left, 1 right move
brown first	**LRL**	1 left, 1 right, 1 left move

Here are the 8 moves for four frogs, two on each side.

green first	**RLLRRLLR**	1 right, 2 left, 2 right, 2 left, 1 right move
brown first	**LRRLLRRL**	1 left, 2 right, 2 left, 2 right, 1 left move

Here are the 15 moves needed for six frogs, three on each side.

green first	**RLLRRRLLLRRRLLR**	1 right, 2 left, 3 right, 3 left, 3 right, 2 left, 1 right move
brown first	**LRRLLLRRRLLLRRL**	1 left, 2 right, 3 left, 3 right, 3 left, 2 right, 1 left move

Use this sequence of moves for 4 frogs per side. Students should used their own **communication** skills to describe the number patterns that they see.

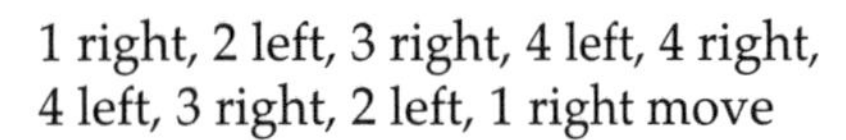

1 right, 2 left, 3 right, 4 left, 4 right, 4 left, 3 right, 2 left, 1 right move

frogs per side	number of moves	number pattern
1	3	1+1+1
2	8	1+2+2+2+1
3	15	1+2+3+3+3+2+1
4	24	1+2+3+4+4+4+3+2+1

STUDENT REACTIONS to Activity 3

Your students may play the game on Worksheet 3 successfully but still not find it easy to complete the chart with the moves they make. These two students offer some useful **problem-solving** advice. Don't try to make the correct moves and record them at the same time. Practice playing a few times first.

play the game serveral times first.

Hint: Play the game several times before you fill in the chart.

This next comment emphasizes the importance of the hands-on part of the activity. Without actually having things to move, the table would be much harder to complete.

It's hard to fill in the chart without the frogs to help

Students like the hops and jumps of the ***Moving Day at Lily Pad Pond*** game and will be eager to play. However, recording their moves correctly poses quite a different problem. They have to organize, remember, and then record their **thinking strategy**.

This activity requires more than just playing the game. It calls for the recording of successive moves and the recognition of underlying patterns. Allow your students ample time.

Moving Day at Lily Pad Pond can serve as the basis for a full day's lesson. Don't be surprised if you get student reactions such as this one.

I give this Lesson 2 thumbs up and I rate this lesson a 10.

MEASUREMENT

5/8 inch

The tiny greenhouse frog is one of the smallest frogs in North America. It lives in Florida and the West Indies and can be as small as 5/8 inch (1.59 cm) in length.

0 1 2 3 4 5

actual size in inches

Introduction to Activity 4

COMPUTING CHANCES
Using Tree Diagrams to Find Probabilities

Description

There are bugs everywhere, and frogs love to catch and eat them for lunch. Different choices of meals are listed and counted in introducing the concept of *probability*. Students do their counting by reading and interpreting results through the use of tree diagrams.

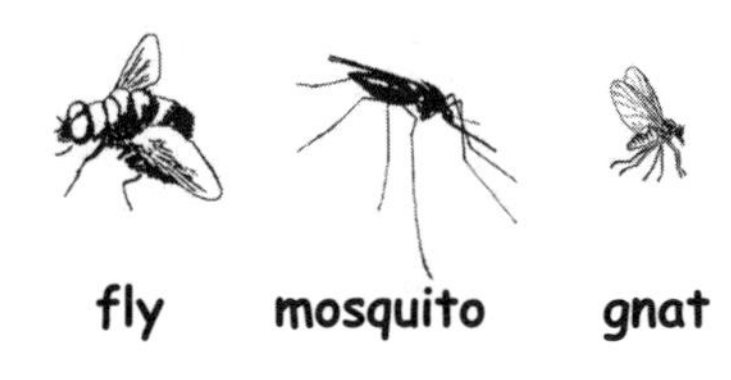

The activity has students count all the different choices defined by a specific event and compare that number to the total number of all possible outcomes to find the chance of the event occurring. That probability or chance can be expressed as a fraction or as a percent.

Big Idea

Tree diagrams are graphs that show lists in a visual, organized, pictorial form. The letters in the different paths or branches of the trees represent the different choices that can be made. We assume here that each choice is equally likely to be made.

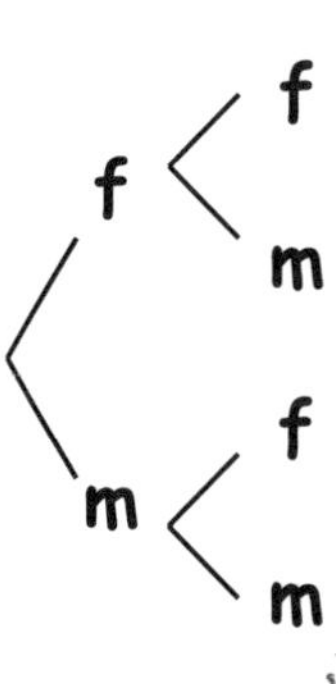

This *tree diagram* shows the 4 choices of 2 bugs.

For a frog, any of these choices is a tasty 2-bug lunch.

Expected Outcomes

Students will gain experience in

- reading and using tree diagrams
- visualizing paths
- counting and comparing results
- computing probabilities
- ordering and arranging

Meeting these NCTM Standards

✓ Numbers
Algebra
✓ Geometry
Measurement
✓ Data Analysis
✓ Problem Solving
Reasoning and Proof
✓ Communication
✓ Connections
✓ Representation

Activity 4 PLANNING HOPS AND JUMPS

Frogs love bugs. They catch them with their tongues and swallow them whole since they have no teeth for chewing.

The Bugs

Q: *A hungry frog eyes three bugs. What are they?*

A: a fly, a mosquito, and a gnat

The Choices

Q: *The frog catches and eats two for lunch. What are the choices?*

A: There are 3 choices. **f** and **m** **f** and **g** **m** and **g**

fly **f**
mosquito **m**
gnat **g**

The Chances

Q: *What are the chances that the frog eats a mosquito?*

A: There are 2 choices with a mosquito.
There are 3 choices in all.
So the chances are 2 out of 3. The answer can be written as the fraction 2/3.

The Problem

Now suppose there are lots of mosquitoes, lots of flies, and lots of gnats. The frog still catches and eats two for lunch, but now they might be two of the same kind of bug.

What are the chances that the frog eats a mosquito?

The Solution

A **tree diagram** or a list can be used to show all the choices, but now keep track of which bug is eaten first and which second.

5 choices contain mosquitoes.
9 choices are possible in all.

So the chances that the frog eats a mosquito are ***5 out of 9*** or ***5/9***.

First bug	Second bug	Choices	Includes a mosquito?
f	f	fly fly	no
	m	fly mosquito	yes ✔
	g	fly gnat	no
m	f	mosquito fly	yes ✔
	m	mosquito mosquito	yes ✔
	g	mosquito gnat	yes ✔
g	f	gnat fly	no
	m	gnat mosquito	yes ✔
	g	gnat gnat	no

Another approach is to make a systematic listing of all choices using the coded letters, **f**, **m**, and **g**, to represent the three bugs. Then count from that list.

ff (fm) fg (mf) (mm) (mg) gf (gm) gg

The order of the letters is important here just as it was with the tree diagram. The first letter identifies the first bug eaten, and the second letter, the second bug.

The Key Ideas

These two key ideas come directly from the subject matter of discrete math.

- The chance that an event occurs is called a *probability*. The probability is found by comparing the number of successful choices to the total number of possible choices. All choices are assumed to be equally likely.

  The probability that the frog eats a mosquito for lunch is ***5 out of 9*** *or* ***5/9***.

- A **tree diagram** is a special graph that shows a systematic listing of outcomes. It is exactly like a regular, organized list except that it is in pictorial form. To use a tree diagram, follow the separate branches from the initial starting point on the left. Each path traces out, in order, a specific choice.

The probability questions on the two activity worksheets have tree diagrams already drawn on them. The important skill developed here is the ability to correctly read and use a tree diagram.

The Extension

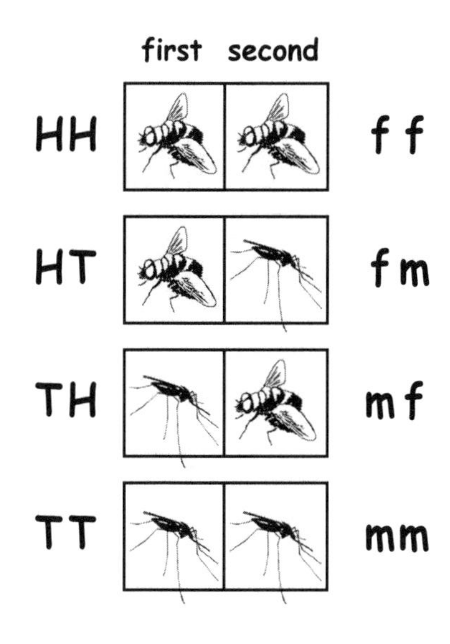

Consider modeling a simple situation with a frog and just two bugs using an experiment. Students identify what bugs are eaten by the toss of two coins. Use a penny for the first bug eaten and a dime for the second. Use the **representation** H, heads, for a fly and T, tails, for a mosquito.

Students, in pairs, toss a penny and a dime 12 separate times. On each toss they record the choices of bugs for lunch. At the end, they count the number that contain a mosquito (tails) and compare that number to the total of 12 tosses.

See how close their experimental probability comes to the answer, *9 out of 12*. Expressed as a fraction, the probability is 9/12 or 3/4.

CATCHING BUGS

Worksheet 4A

It's lunch time. A frog is looking for a meal.
There are lots of flies and lots of mosquitoes to choose from.
The frog catches and eats two of them.

1. Use the tree diagram to help you complete the table.
   List the choices using **f** for a fly and **m** for a mosquito.
   Then answer **yes** or **no** to the question.

TREE DIAGRAM

First bug	Second bug	Choices	Includes a mosquito?
f	f	___ ___	________
	m	___ ___	________
m	f	___ ___	________
	m	___ ___	________

2. How many of the choices contain a mosquito? _____

   How many different choices are there in all? _____

3. The frog eats two bugs for lunch. What are the chances that one or two of them are mosquitoes? ____ out of ____

   Write the probability as a fraction. _____

4. What are the chances that both bugs are alike, either two flies or two mosquitoes? ____ out of ____

MORE CHOICES

Worksheet 4B

Frogs love bugs. They catch them with their tongues and swallow them whole since they have no teeth for chewing. It's dinner time and lots of flies, mosquitoes, and gnats are around just waiting for the frog to catch and eat them.

1. Complete the table.
   Use **f** for fly, **m** for mosquito, and **g** for gnat.

TREE DIAGRAM

First bug	Second bug	Choices	Includes a mosquito?
f	f	___ ___	___
	m	___ ___	___
	g	___ ___	___
m	f	___ ___	___
	m	___ ___	___
	g	___ ___	___
g	f	___ ___	___
	m	___ ___	___
	g	___ ___	___

2. How many of the choices contain a mosquito? ___ ___
   How many different choices are there in all?
   What are the chances that a mosquito is eaten? ___ out of ___

3. What are the chances that only flies are eaten? ___ out of ___

4. What are the chances that the bugs are different? ___ out of ___

5. Write the three probabilities above as fractions. ___ ___ ___

6. What is the probability that only the second bug is a gnat? ___

ANSWERS to Worksheets 4A and 4B

CATCHING BUGS — Worksheet 4A

It's lunch time. A frog is looking for a meal.
There are lots of flies and lots of mosquitoes to choose from.
The frog catches and eats two of them.

1. Use the tree diagram to help you complete the table.
   List the choices using **f** for a fly and **m** for a mosquito.
   Then answer **yes** or **no** to the question.

TREE DIAGRAM

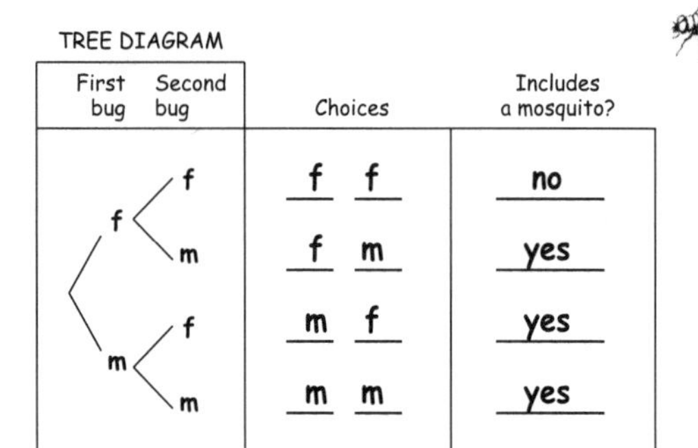

First bug	Second bug	Choices	Includes a mosquito?
f	f	f f	no
	m	f m	yes
m	f	m f	yes
	m	m m	yes

2. How many of the choices contain a mosquito? **3**

   How many different choices are there in all? **4**

3. The frog eats two bugs for lunch. What are the chances that one or two of them are mosquitoes? **3** out of **4**

   Write the probability as a fraction. **3/4**

4. What are the chances that both bugs are alike, either two flies or two mosquitoes? **2** out of **4**

46

MORE CHOICES — Worksheet 4B

Frogs love bugs. They catch them with their tongues and swallow them whole since they have no teeth for chewing.
It's dinner time and lots of flies, mosquitoes, and gnats are around just waiting for the frog to catch and eat them.

1. Complete the table.
   Use **f** for fly, **m** for mosquito, and **g** for gnat.

TREE DIAGRAM

First bug	Second bug	Choices	Includes a mosquito?
f	f	f f	no
	m	f m	yes
	g	f g	no
m	f	m f	yes
	m	m m	yes
	g	m g	yes
g	f	g f	no
	m	g m	yes
	g	g g	no

2. How many of the choices contain a mosquito? **5**
   How many different choices are there in all? **9**
   What are the chances that a mosquito is eaten? **5** out of **9**
3. What are the chances that only flies are eaten? **1** out of **9**
4. What are the chances that the bugs are different? **6** out of **9**
5. Write the three probabilities above as fractions. **5/9 1/9 6/9**
6. What is the probability that only the second bug is a gnat? **2/9**

47

FROG Facts

A frog's diet might consist of insects, but some frogs eat small mammals and even other frogs. The mosquitoes they eat help control malaria and many other diseases transmitted by these bugs. Frogs are a favorite diet of bats, snakes, and other animals.

TRANSPARENCY MASTERS

Make transparencies from these masters.

Use the left hand figure on this transparency when discussing the tree diagram for Worksheet 4A. Mask out the diagram that is on the right. Trace out the four different branches of the tree on the left with your finger. They are *ff*, *fm*, *mf*, and *mm*. This will help students see how these four different choices of bugs are read from this tree diagram.

The figure on the right illustrates how a tree diagram would look if the frog ate three bugs for lunch. On this tree, each of the eight branches contains three letters. These represent the eight different choices of flies and/or mosquitoes. They are *fff*, *ffm*, *fmf*, *fmm*, *mff*, *mfm*, *mmf*, and *mmm*.

Use this transparency in conjunction with a discussion of Worksheet 4B. The nine different branches of this tree represent the nine different ways the frog can catch a fly or a mosquito or a gnat for each of its two bugs.

All choices on the branches of a given tree are equally likely to occur. What this assumes, in effect, is that the frog is just as likely to catch a fly as a mosquito. This allows us to count and compare choices as we do.

In the real world, for a given frog, a given fly, and a given mosquito, that condition of equality may not be met. Discuss with your students whether they think it is easier for a frog to catch a fly or a mosquito, or if the task is of about the same difficulty for each bug.

BUGS FOR LUNCH

fly

mosquito

Choices for a 2-bug lunch

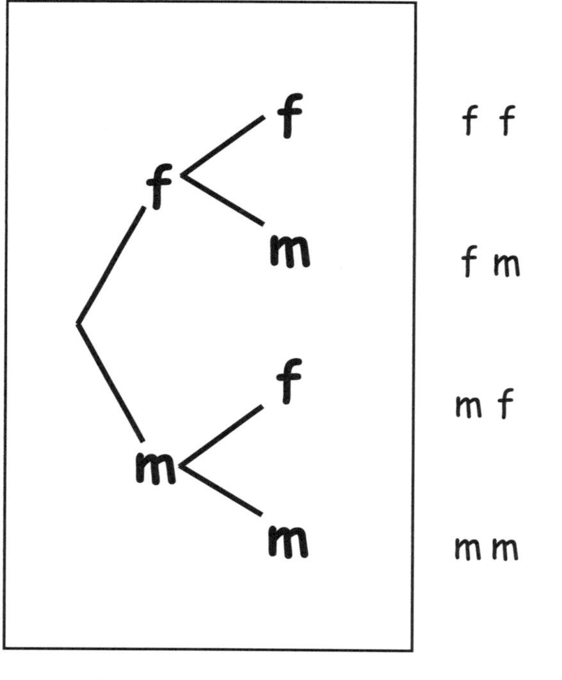

TREE DIAGRAM

Choices

f f

f m

m f

m m

Choices for a 3-bug lunch

First bug

Second bug

Third bug

f
f
f
m
m
f
m
m
f
f
m
m
f
m

TREE DIAGRAM

Choices

f f f

f f m

f m f

f m m

m f f

m f m

m m f

m m m

A

FLIES, MOSQUITOES, and GNATS

Choices for a 2-bug lunch

First bug Second bug

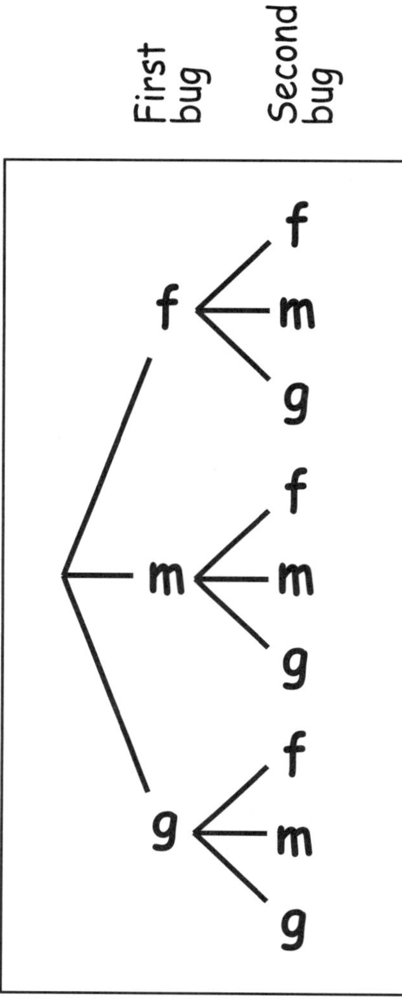

fly	fly
fly	mosquito
fly	gnat
mosquito	fly
mosquito	mosquito
mosquito	gnat
gnat	fly
gnat	mosquito
gnat	gnat

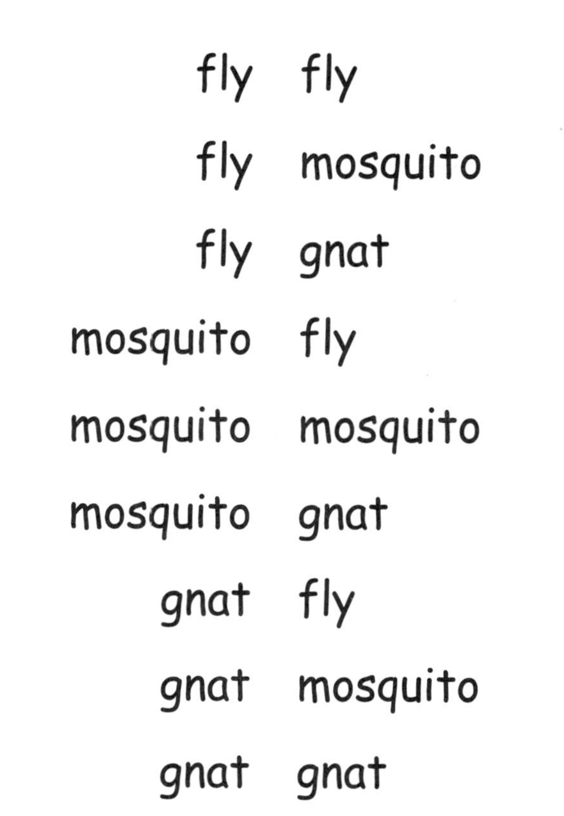

B

GETTING THE MOST from Activity 4

Some students will need special help in matching each path in a **tree diagram** with its corresponding entry in an ordered list. Use transparency A. Trace out different paths on the tree with your finger, and have the class give the corresponding name by reading the letters on that branch in the correct order.

Note in this example that two different paths go through the same f on the left while only one path goes to the first f on the right.

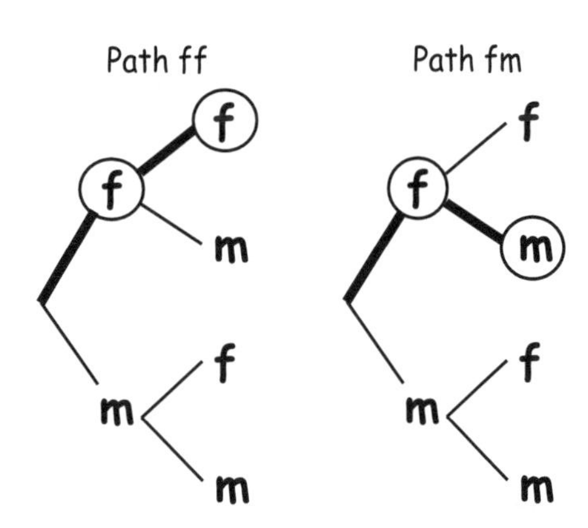

The tree diagram helped me put the list together, but the list helped me more to answer the questins because it's writen down in order, It just made more sense to me,

Many students, such as this one, will be able to construct the tree diagram but will choose to use the list for counting.

The activity on Worksheet 4A can be expanded to include more **problem-solving** situations. Keep the two kinds of bugs but have the frog eat three bugs. The new tree diagram has 8 separate branches.

Here are some probabilities for the frog's lunch of three bugs.

all 3 mosquitoes	**1 out of 8**
exactly 2 mosquitoes	**3 out of 8**
at least 1 mosquito	**7 out of 8**

As the frog eats more and more bugs, the chance or probability that a mosquito is eaten continues to increase.

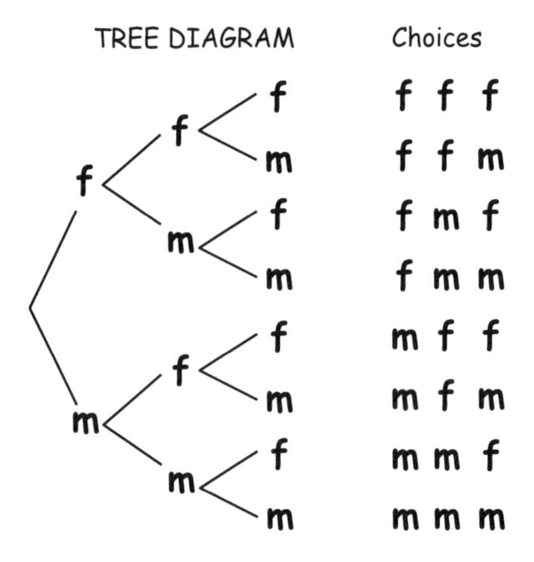

Puzzle C, on page 113, offers a further extension of this activity. It includes a third choice of an insect, a gnat, for the frog's lunch.

STUDENT REACTIONS to Activity 4

Frogs are not the only ones who like bugs. You'll find students love them too, but in a very different way.

Here are some students' reflections on using the tree diagram and the list for counting. This kind of written **communication** by the student can be very revealing.

Choices		Includes a mosquito?
F	F	no
F	m	yes
F	g	no
M	F	yes
M	M	yes
M	g	yes
G	F	No
G	m	yes
G	G	no

The list was easier because it was easier to write down

The tree diagram helped me put the awensers in order.

The tree diagram helped me make the list but the list was easier.

This remark is from a student who believes that a fraction is the required form for expressing a probability.

In order to answer probabilitys you have to know your fractions.

5 out of 9
1 out of 9
6 out of 9

$\frac{5}{9}$ $\frac{1}{9}$ $\frac{6}{9}$

Different forms of **representation** can be used to express the same probability. Your choices should be appropriate to the grade level of the students involved. Here are some different forms for expressing the same probability.

6 out of 9 6/9 2 out of 3 two-thirds 66 2/3% 0.666...

Show your students the **connection** between probabilities and points on the number line between 0 and 1. The nearer the probability is to 1, the more certain are the chances of the event occurring. A probability of 1 means the event must occur. A probability of 0 means it cannot occur.

MATH and FROGS

PERCENT

This is a spotted dart-poison frog from Venezuela. Would 50%, 60%, 70%, or 80% be the best estimate of the percent of its body surface that is black?

Introduction to Activity 

DRAWING PICTURES
Listing Arrangements as Pictures

Description

Different arrangements can be listed using letters, numbers, or other symbols. These same arrangements can be presented as networks of paths through a given set of vertices. This simple process can lead to many interesting applications of discrete mathematics using graphs.

Students work with a single set of five vertices. They trace all possible paths that start at a given vertex, travel through the other vertices, and return to the starting vertex. These different paths are shown in the form of graphs representing the different ways a frog can hop among lettered lily pads. Direct applications can be made to many real world problems.

Start G F O R

RGFO

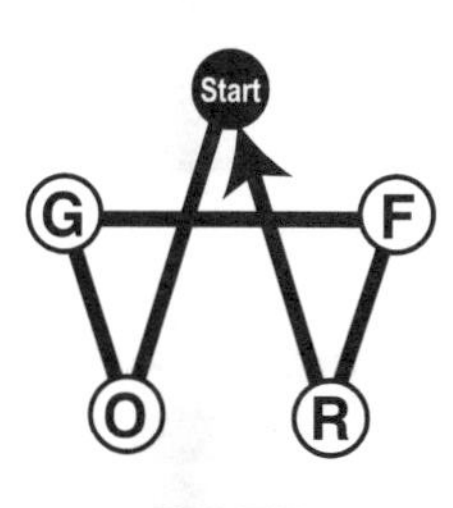

OGFR

Trace through the same set of vertices but in a different order and you get a different picture.

Big Idea

A **network** is a special kind of graph. It can be thought of as a picture of vertices connected by edges. In this activity, the vertices are letters with the edges forming a path connecting the letters in some special order.

Expected Outcomes

Students will gain experience in

- tracing paths
- listing systematically
- representing arrangements as networks
- visualizing symmetries and rotations
- searching for patterns

Meeting these NCTM Standards

Numbers	✓ Problem Solving
Algebra	Reasoning and Proof
✓ Geometry	✓ Communication
Measurement	✓ Connections
✓ Data Analysis	✓ Representation

DRAWING PICTURES

Getting Ready

Begin with a list of the 24 different arrangements of the four letters in the word FROG. Review the counting procedure that was used.

FGOR	GFOR	OFGR	RFGO
FGRO	GFRO	OFRG	RFOG
FOGR	GOFR	OGFR	RGFO
FORG	GORF	OGRF	RGOF
FRGO	GRFO	ORFG	ROFG
FROG	GROF	ORGF	ROGF

There are 4 choices for the first letter, 3 remaining for the second, 2 for the third, and only 1 left for the fourth. Multiply to get 24, the total number.

$$4! = 4 \times 3 \times 2 \times 1 = 24$$

Hopping Around

A frog hops to each of the four lettered lily pads, visiting each one just once, and then returns. There are 24 possible paths. Each path forms a **network** showing one of the 24 arrangements listed above.

Picturing the Arrangements

Use lettered circles to represent the lily pads. Start and end at the top. Trace a path that goes through each of the four letters.

The path for RFGO looks something like the letter W. Reverse the last two letters for RFOG and you get a very different **representation**.

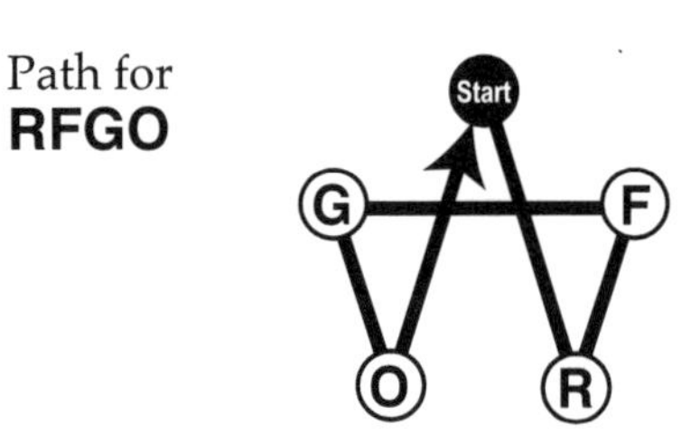

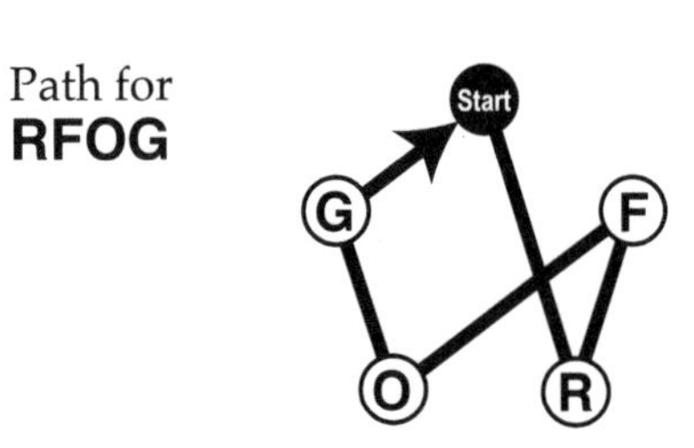

Posing the Problems

Find two paths that start and end at the top and form a polygon. Name the polygon.

How many different paths can you find that form 5-pointed stars?

How many different paths have the same shape as RFGO? Be sure to count rotations.

Solutions

These two networks show the same polygon, a pentagon, where the only difference is the direction traveled. The first path, FROG, goes clockwise, while the second path, GORF, goes counterclockwise.

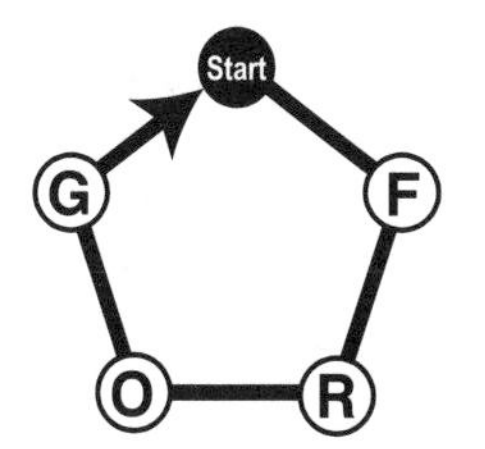

Path for **FROG**

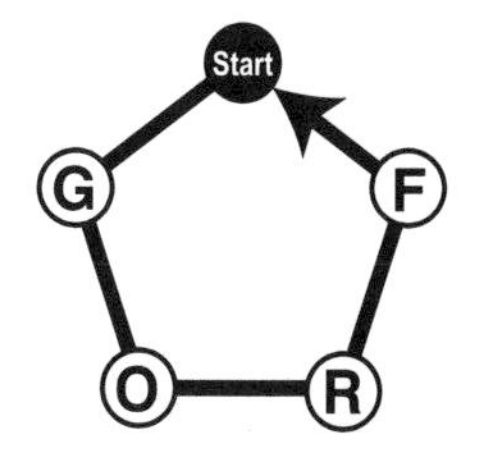

Path for **GORF**

The same **reasoning** tells us there must be two 5-pointed stars, with one the reverse order of the other.

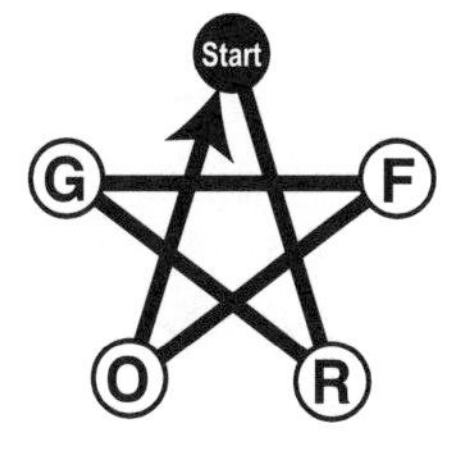

Path for **RGFO**

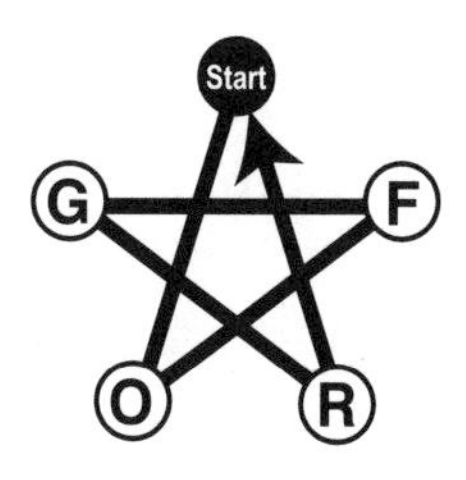

Path for **OFGR**

The network shown below is the graph for RFGO. It looks a bit like the letter W. There are 5 different rotations possible with this shape. When the direction is reversed, there are another 5.

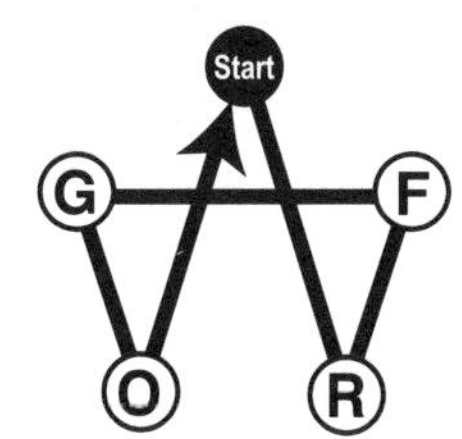

Path for **RFGO**

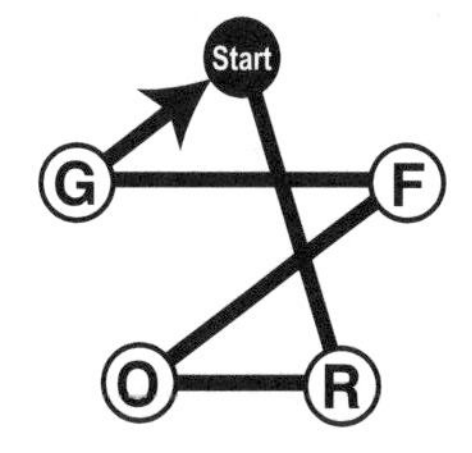

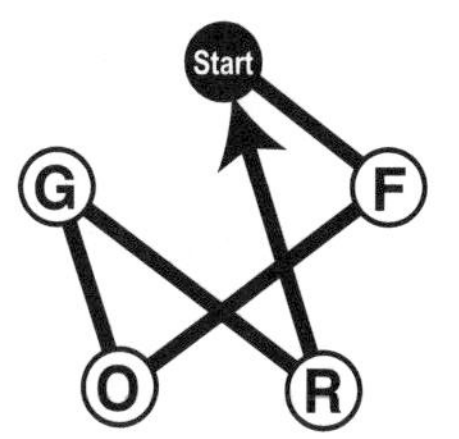

Let your students verify the **connection** between these 10 lettered paths and this single basic geometric shape.

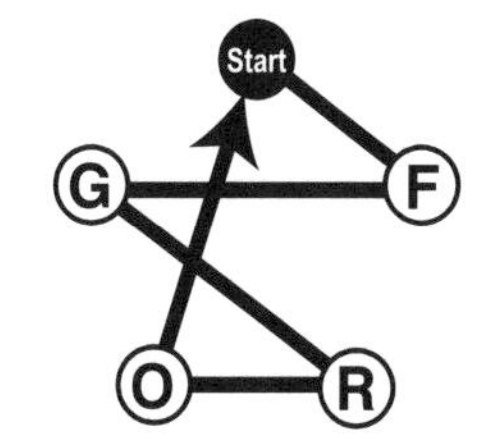

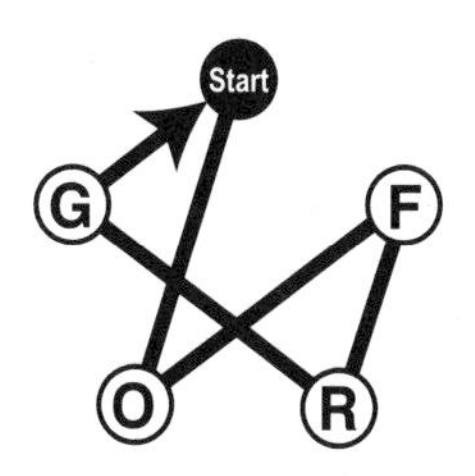

RFGO OGFR
ROFG GFOR
FOGR RGOF
OFRG GRFO
FGRO ORGF

Extension

Have your students search for other paths that have different shapes. In all, only 4 shapes are possible that start and end at the top and trace through each of the 4 letters only once. There are 2 pentagons, 2 stars, 10 shaped like a W as shown above, and another 10 like the picture of RFOG on the right.

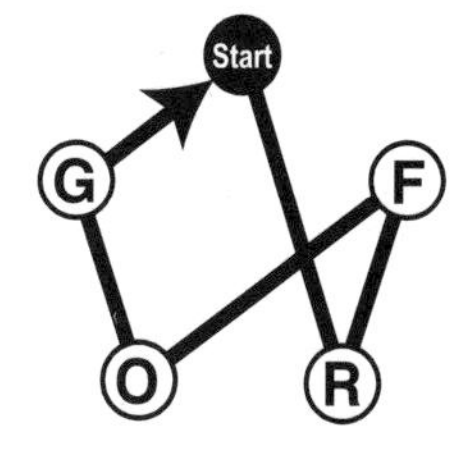

Path for **RFOG**

HOPPING TO AND FRO

This frog has a problem, a math problem. It wants to hop to the other four lily pads, visiting each one just once, before returning.

Only the one path shown goes through the lily pads in alphabetic order. However, the frog can't read the letters.

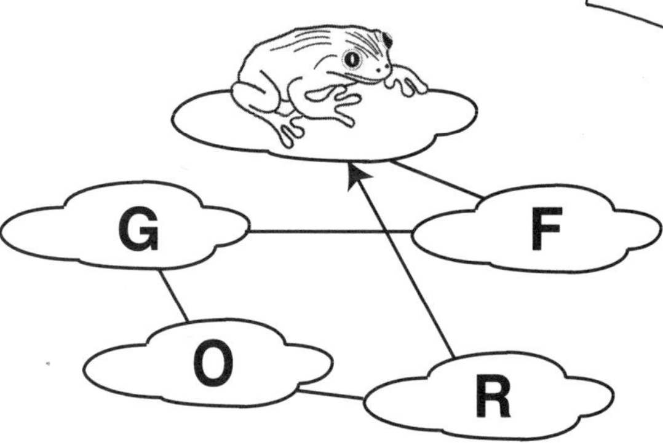

1. Use this diagram to show the path through the letters in alphabetic order. Remember, start at the top and end there as well. Use an arrowhead at the end to indicate the direction.

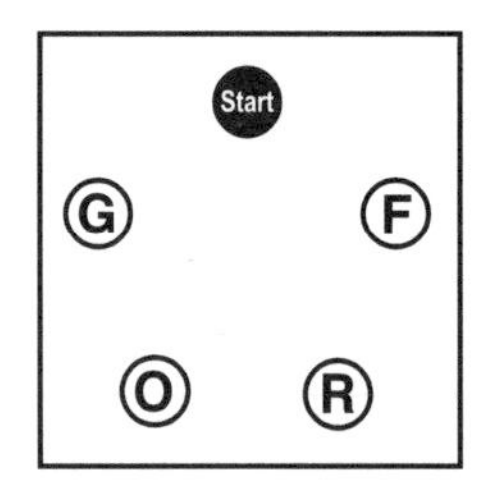

2. The frog sees two paths that are polygons. Use the letters to name these two polygons. What kind of polygon does the frog see?

3. The frog starts at the top, hops to each letter just once, and ends at the top again, but it never hops to an adjacent lily pad. Draw the two possible choices on these diagrams. How would you describe their shapes?

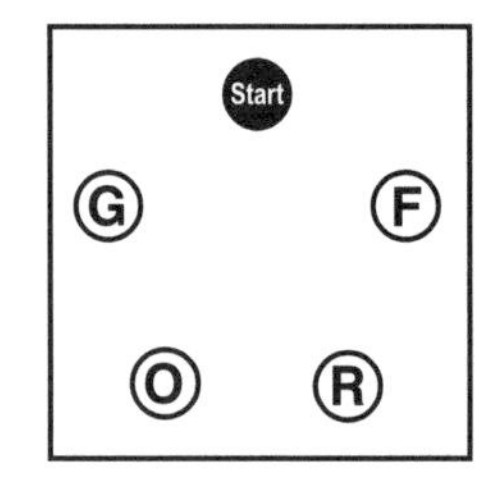

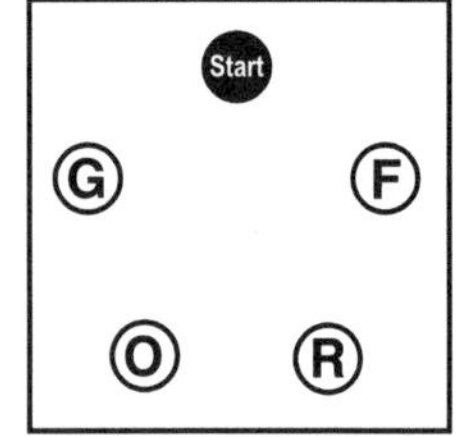

4. Draw four more paths that have the same shape as the path for RFGO, each at a different rotation. Give the names of these paths.

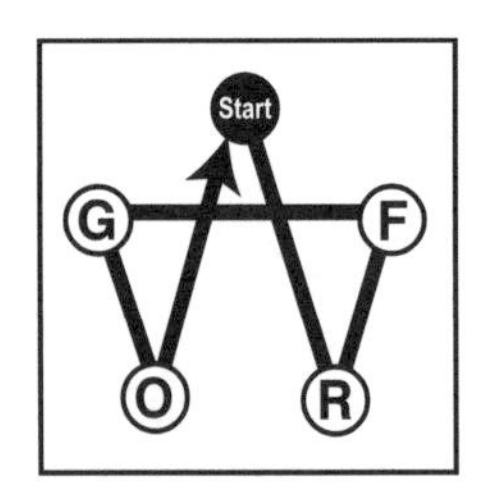

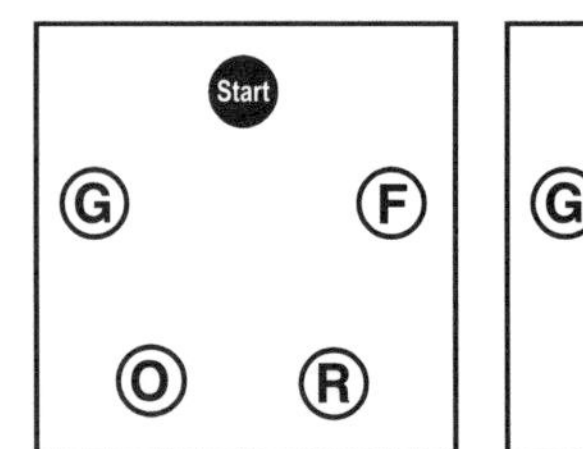

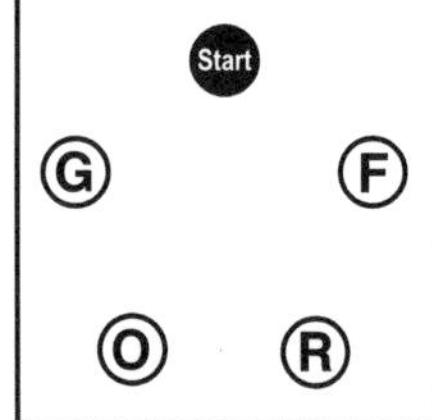

5. The 6 paths that go to F first are already drawn. Draw the 6 paths that go to R first. Then draw those that start with O and those that start with G.

Go to F first.	Go to R first.	Go to O first.	Go to G first.
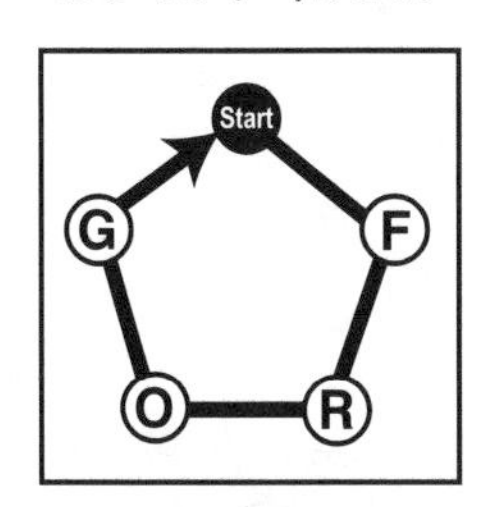	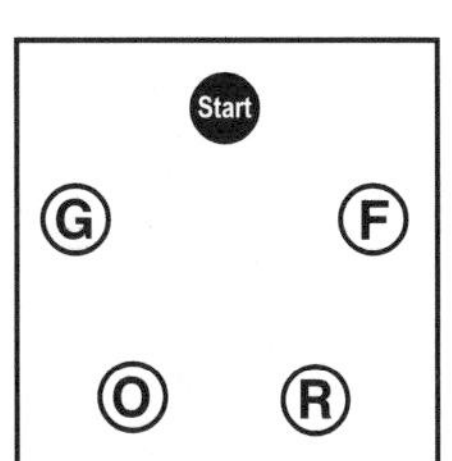	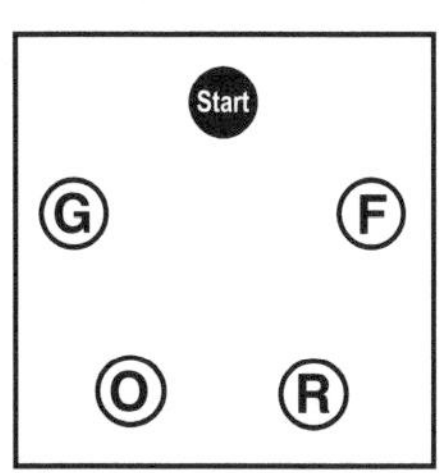	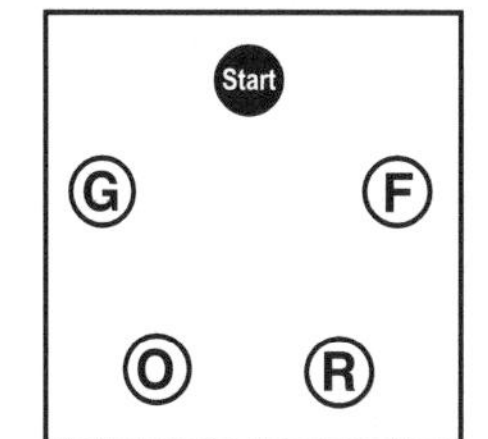
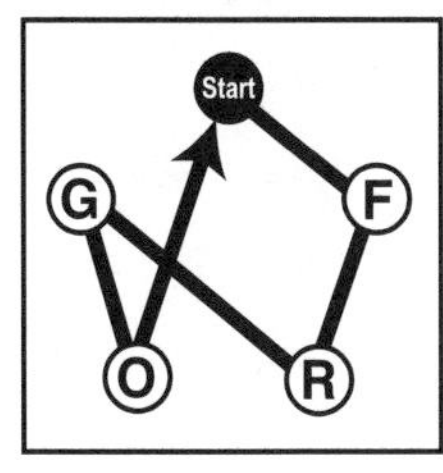	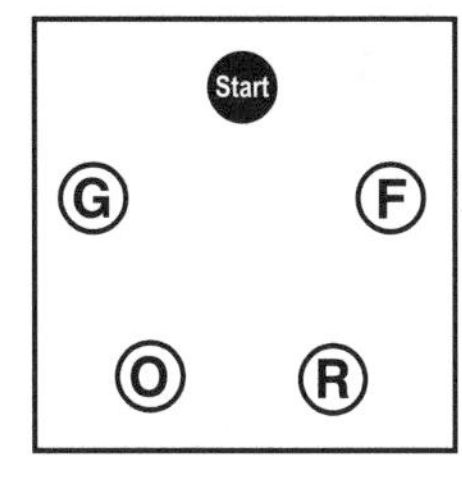	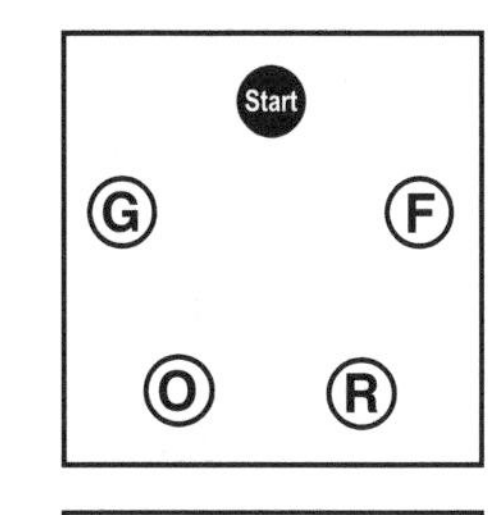	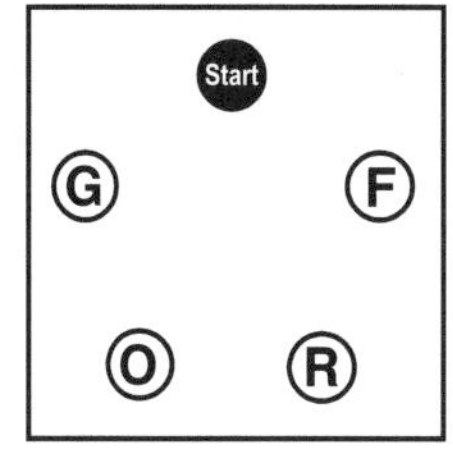
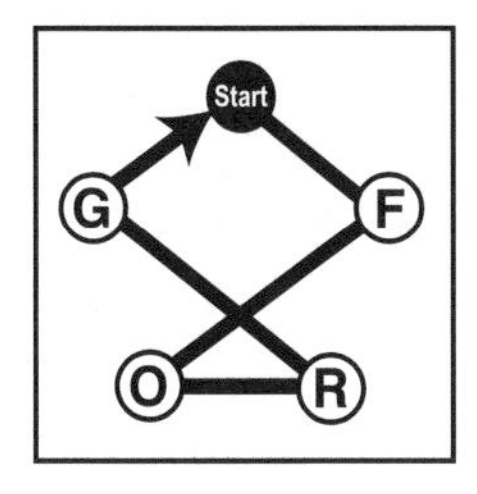	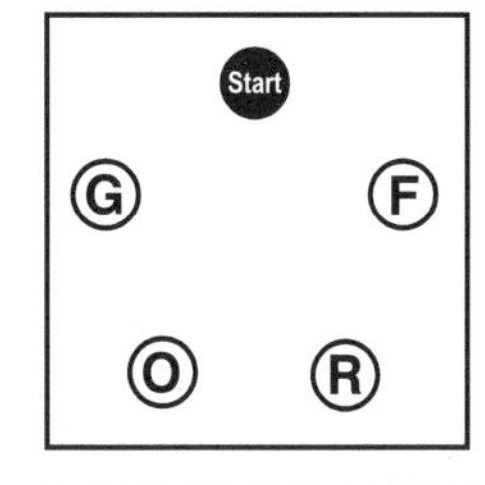	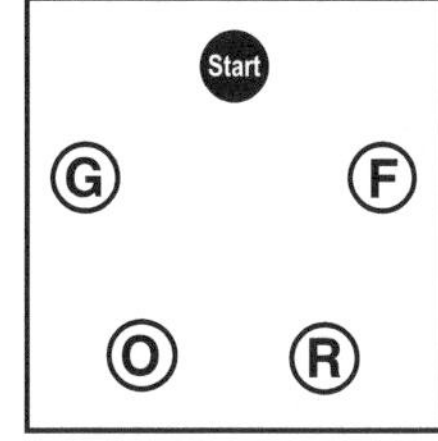	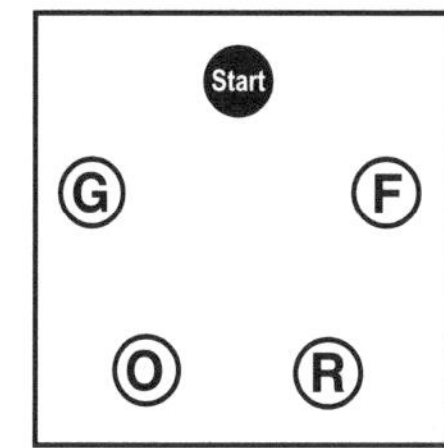
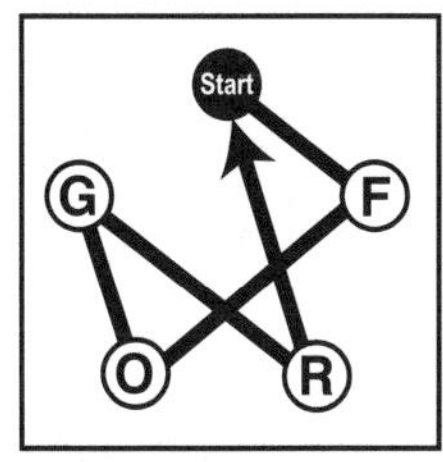	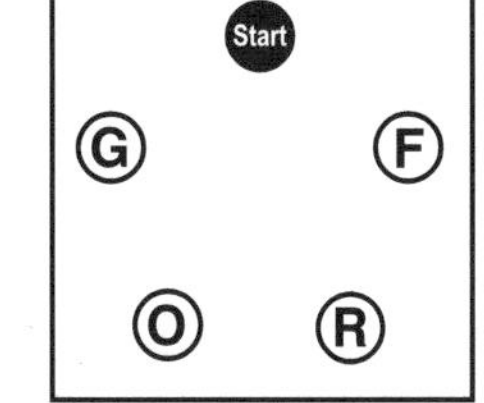	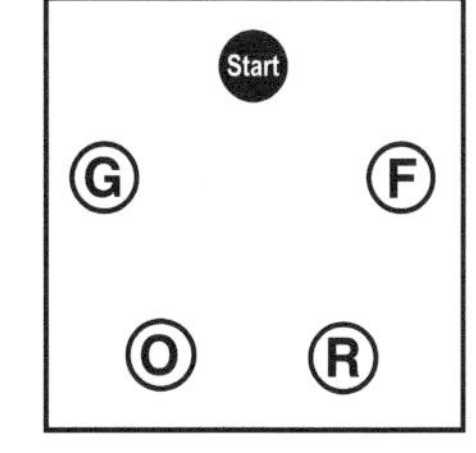	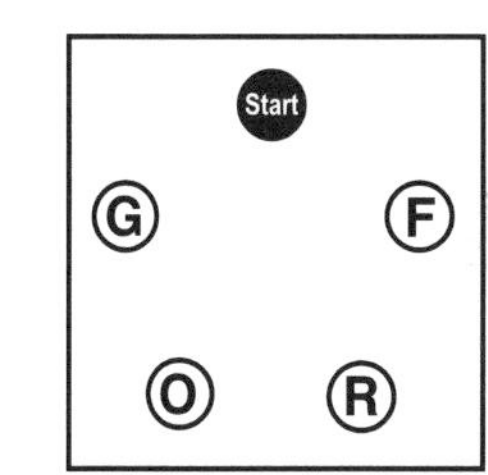
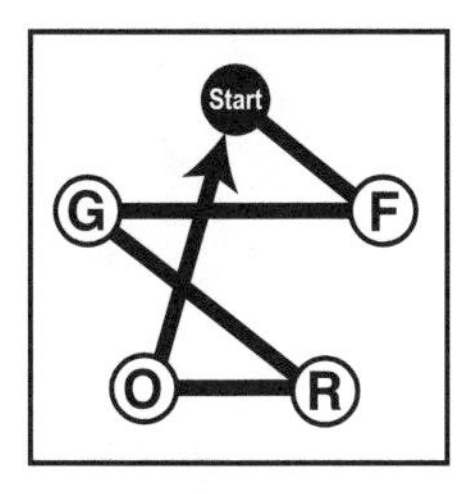	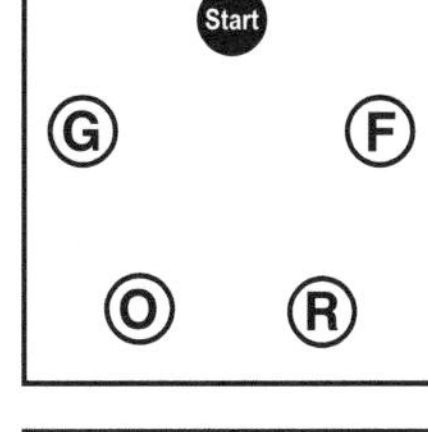	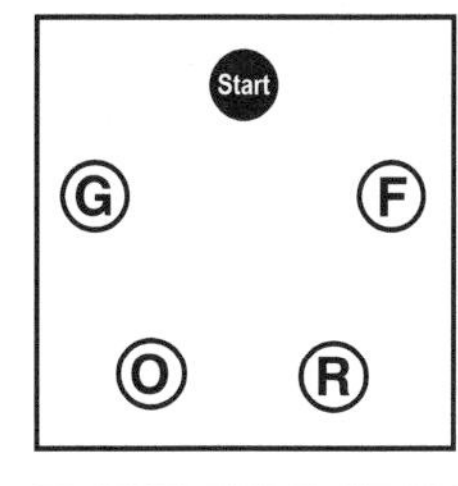	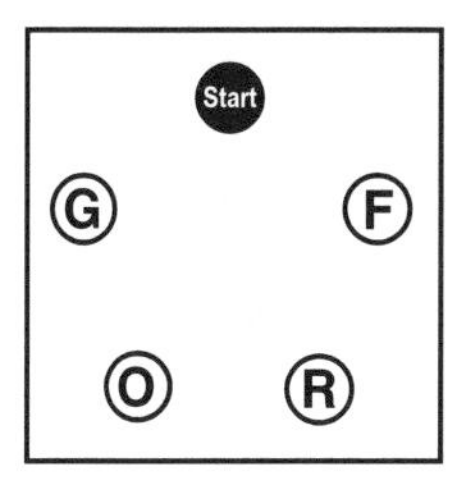
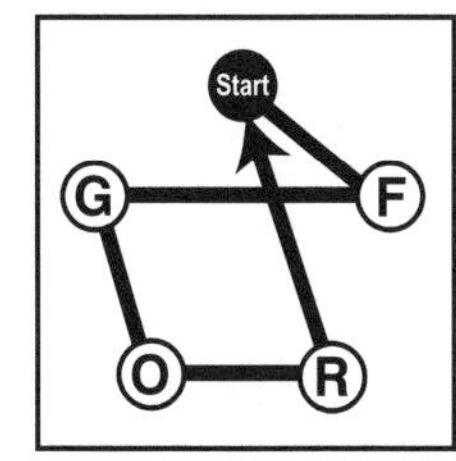	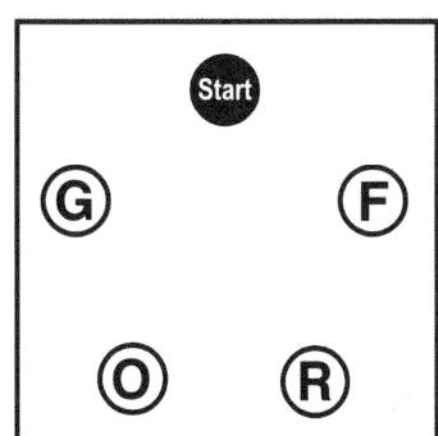	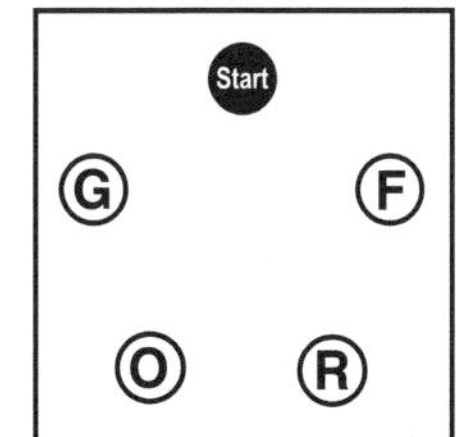	

ANSWERS to Worksheets 5A and 5B

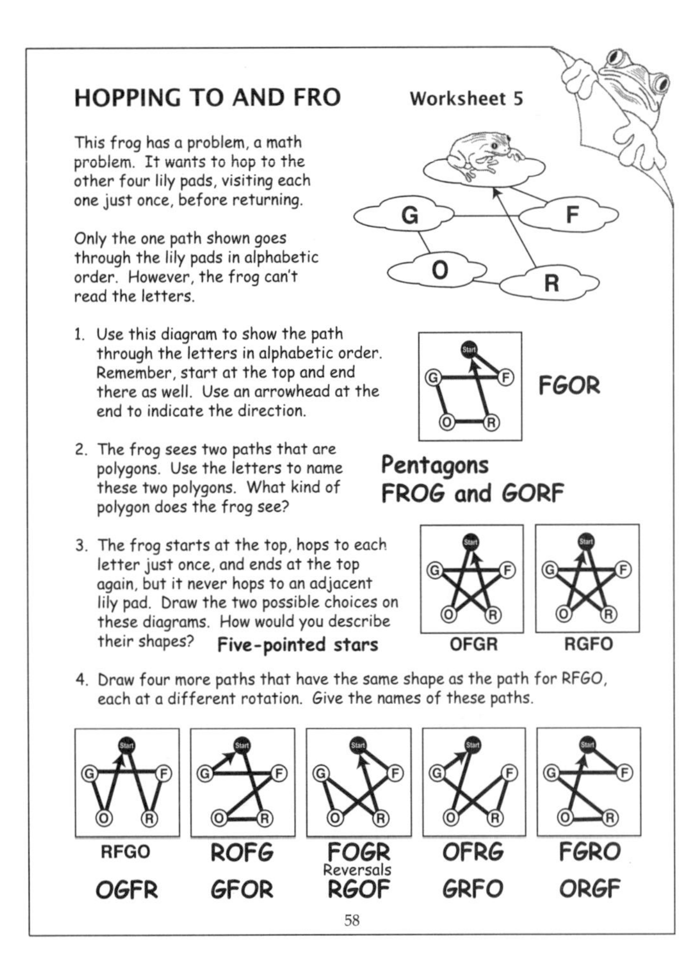

HOPPING TO AND FRO Worksheet 5

This frog has a problem, a math problem. It wants to hop to the other four lily pads, visiting each one just once, before returning.

Only the one path shown goes through the lily pads in alphabetic order. However, the frog can't read the letters.

1. Use this diagram to show the path through the letters in alphabetic order. Remember, start at the top and end there as well. Use an arrowhead at the end to indicate the direction. **FGOR**
2. The frog sees two paths that are polygons. Use the letters to name these two polygons. What kind of polygon does the frog see? **Pentagons FROG and GORF**
3. The frog starts at the top, hops to each letter just once, and ends at the top again, but it never hops to an adjacent lily pad. Draw the two possible choices on these diagrams. How would you describe their shapes? **Five-pointed stars** **OFGR** **RGFO**
4. Draw four more paths that have the same shape as the path for RFGO, each at a different rotation. Give the names of these paths.

RFGO	ROFG	FOGR	OFRG	FGRO
		Reversals		
OGFR	**GFOR**	**RGOF**	**GRFO**	**ORGF**

58

5. The 6 paths that go to F first are already drawn. Draw the 6 paths that go to R first. Then draw those that start with O and those that start with G.

Go to F first. Go to R first. Go to O first. Go to G first.

59

FROG Facts

Amphibian is a class of animals with backbones. They are cold-blooded animals, so their body temperatures vary with their surroundings. Frogs are amphibians. Most frogs lay their eggs in water, but they can live both on land and in the water.

am•phib•i•an

TRANSPARENCY MASTERS

Make transparencies from these masters.

Students will want to check the paths they create for question 5 on the worksheet. This transparency contains two solution sets. The first set, on the left, is based on the method used for the examples on Worksheet 5B. The systematic listing is based on the sequence FROG. Successive entries move through these letters in that order.

The second set, on the right, contains the same 24 paths, but their order is different. These arrangements are in alphabetic order, starting with the sequence FGOR and ending with ROGF.

Some students will want the challenge of creating the entire list themselves. Others will need all the help they can get. For these latter students, alphabetizing may be the more friendly process to follow. Either way, the ability to create lists in a systematic fashion remains a critical component of the essential problem-solving skills needed for mathematics.

Use transparency B for an activity combining mathematics and science. Have students look up the four major areas where rain forests exist and locate them on this map. They are

- Equatorial Africa
- Tropical South America
- Central America and the West Indies
- Indonesia

Let these be the four vertices, E, T, C, and I, of a graph, where the fifth vertex, H, is the home state where the students live.

Now use the 24 different possible paths from Activity 5 to represent the 24 different ways the students can travel from home to these 4 rain forest areas and return. Let students trace individual paths on the map and discuss which would be the shortest.

Based on the sequence FROG

F first	R first	O first	G first
FROG	RFOG	OFRG	GFRO
FRGO	RFGO	OFGR	GFOR
FORG	ROFG	ORFG	GRFO
FOGR	ROGF	ORGF	GROF
FGRO	RGFO	OGFR	GOFR
FGOR	RGOF	OGRF	GORF

Alphabetized from FGOR

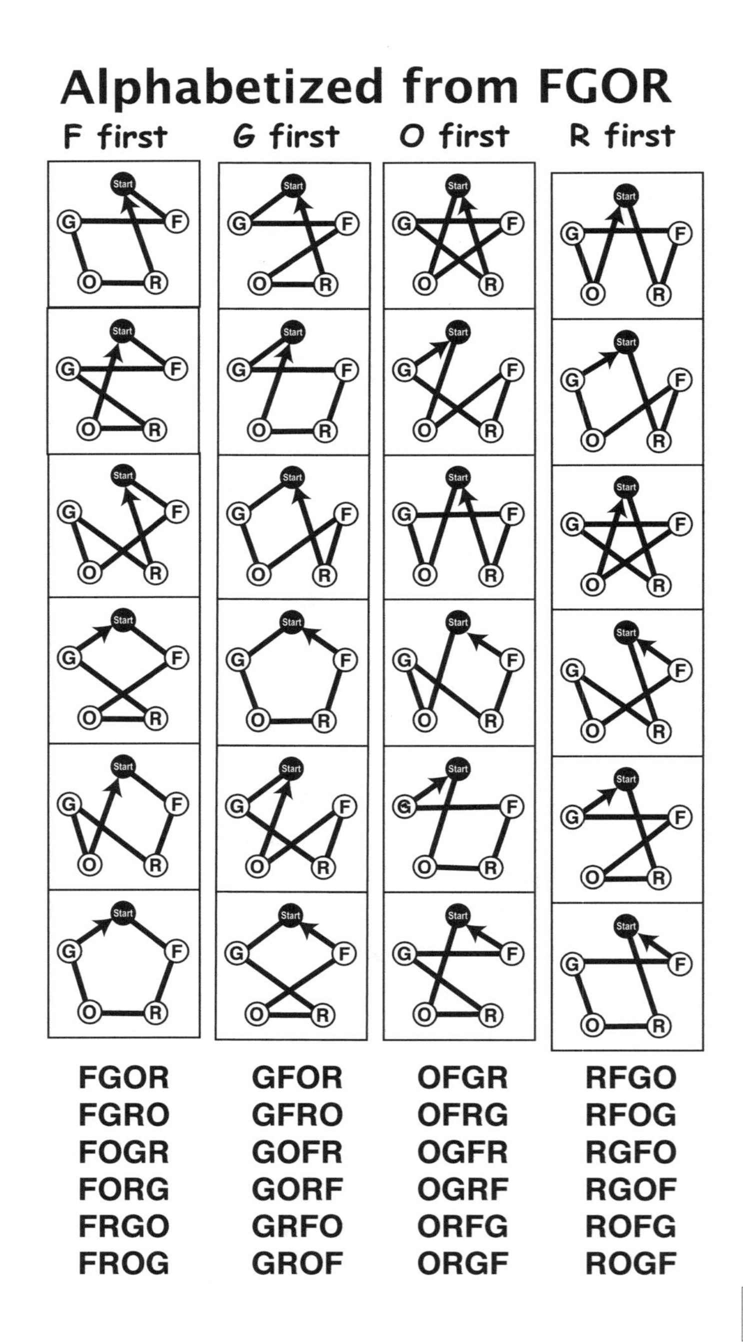

F first	G first	O first	R first
FGOR	GFOR	OFGR	RFGO
FGRO	GFRO	OFRG	RFOG
FOGR	GOFR	OGFR	RGFO
FORG	GORF	OGRF	RGOF
FRGO	GRFO	ORFG	ROFG
FROG	GROF	ORGF	ROGF

A

LOCATE THE FOUR MAJOR TROPICAL RAIN FOREST AREAS

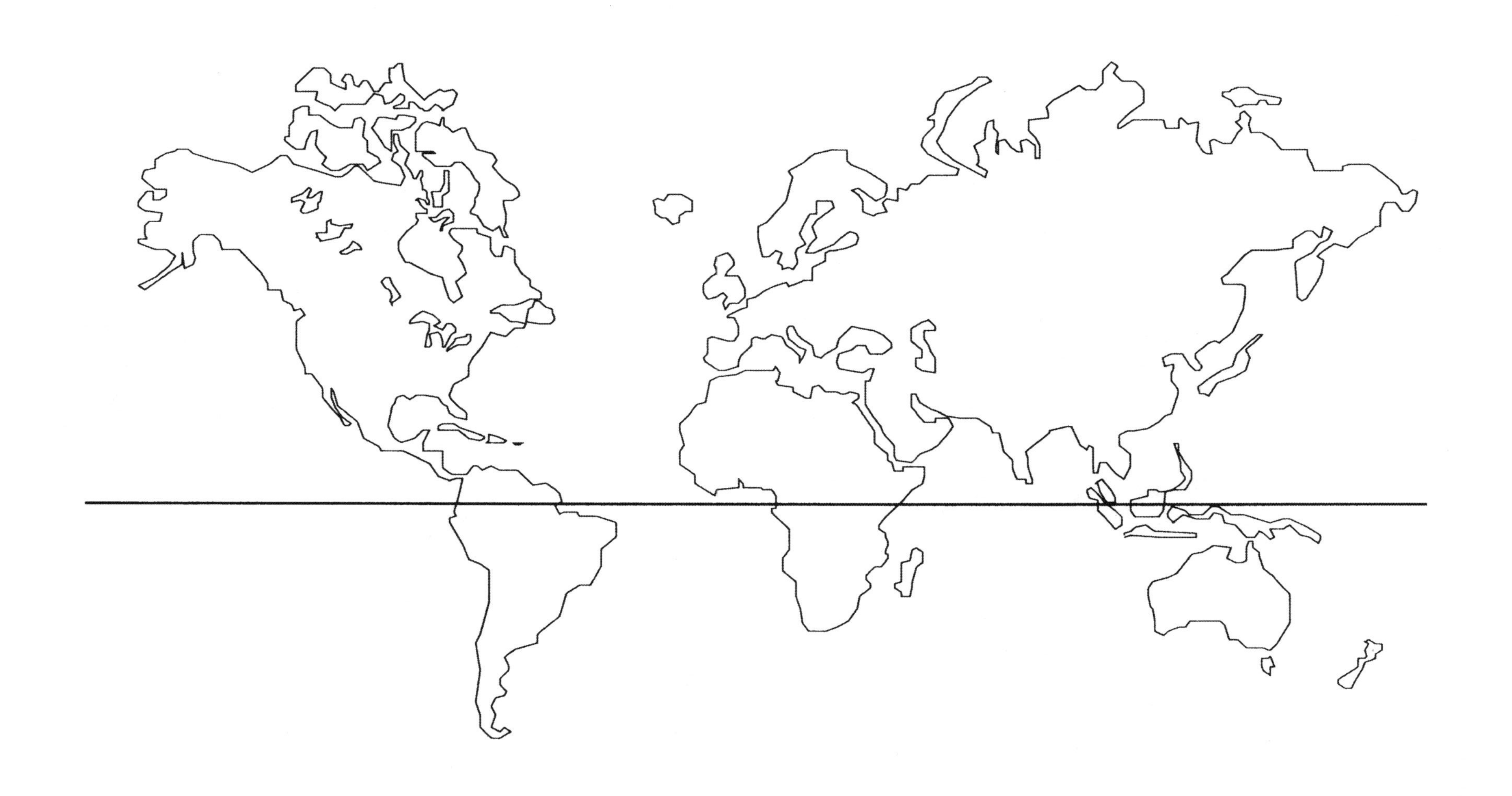

B

GETTING THE MOST from Activity 5

This activity has students explore various paths through a given set of 5 vertices. There are two worksheets for the one activity. Begin question 5 by having students study the six paths shown that go first to the letter F. Then let them try to find all the paths that go first to the letter R, O, and G.

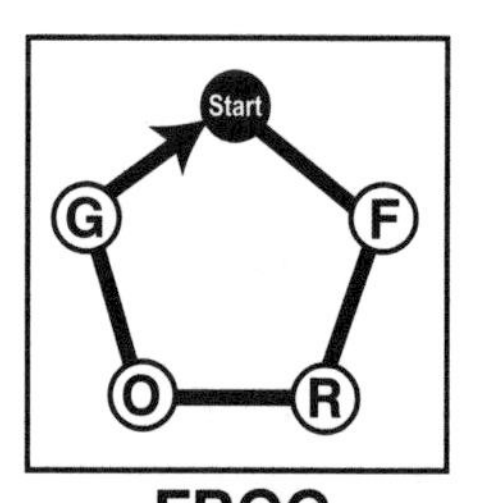

FROG
5 single steps

One of the key NCTM standards in mathematics is **representation**, where different models are used to represent a single idea. These networks or graphs are one way to represent different paths. The lettered sequences are another. A third might involve numerical patterns. Discuss with your students the specific advantages and disadvantages of each form.

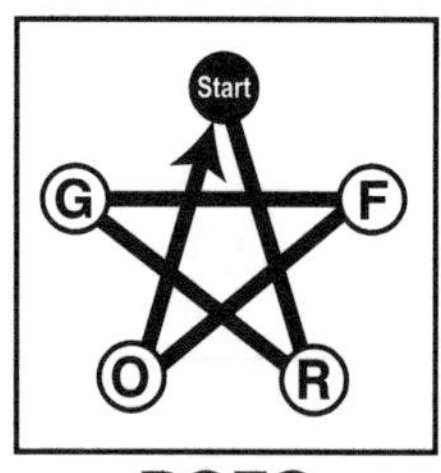

RGFO
5 double steps

You can move from any vertex to any other vertex by taking 1 or 2 steps around the vertices in either direction. This idea can be used to describe each path as a sequence of numbers.

The path **FROG** becomes **11111**. Its length is 5.
The path **RGFO** becomes **22222**. Its length is 10.

The path **OGFR** becomes **21212**. Its length is 8.
The path **GROF** becomes **12121**. Its length is 7.

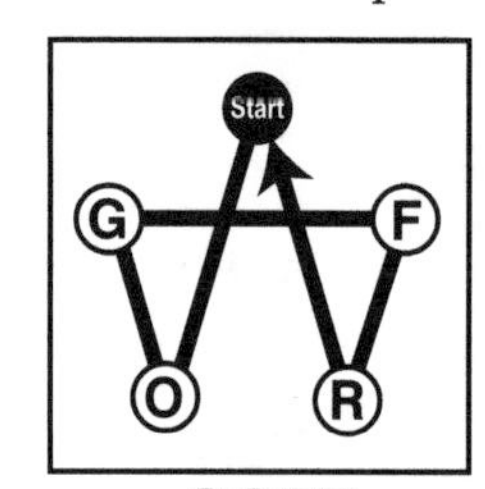

OGFR
2 single and
3 double steps

Students need to see the **connection** between networks and real world applications. Let home be Texas. There are 24 different routes to follow in visiting the 4 states of Florida (**F**), Georgia (**G**), Oregon (**O**), and Rhode Island (**R**) and returning home. The path **FROG**, shown here, represents one of those 24 possible trips.

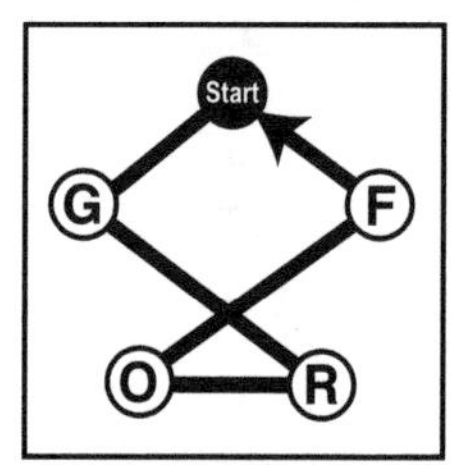

GROF
3 single and
2 double steps

STUDENT REACTIONS to Activity 5

Don't be surprised if your students have trouble completing all 24 paths for question 5 on the worksheet. Organized, systematic listing can be very challenging for many students at many different levels of ability.

Of course, there will always be some students, such as Isabelle, who will question why any solutions at all were given on the worksheet.

Don't give away too many answers.

Take time to discuss with your students the need for an organized way of making a list of all possible arrangements. Good planning and **reasoning** are critical here. Too many students will simply start out drawing paths at random, get about half way through, and then not know how to continue.

Some students will go to a given letter first but not exhaust all the possibilities before trying another first letter. Alex cites yet another possible difficulty.

You might make a double shape. And not know about it.

Here is a listing strategy that is based on the sequence of letters in the word FROG. It is systematic, non-repetitious, and exhaustive.

List 2 paths from FR, 2 from FO, and 2 from FG.
List 2 paths from RF, 2 from RO, and 2 from RG.
List 2 paths from OF, 2 from OR, and 2 from OG.
List 2 paths from GF, 2 from GR, and 2 from GO.

One effective **problem-solving** strategy is to have students identify errors such as those shown in these student samples.

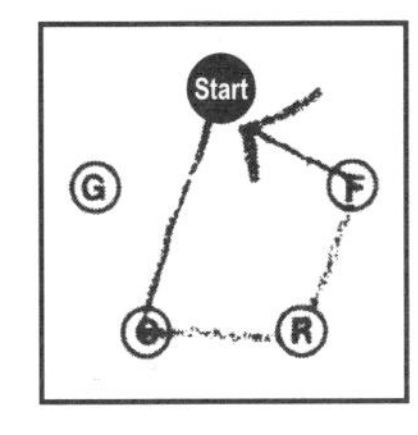

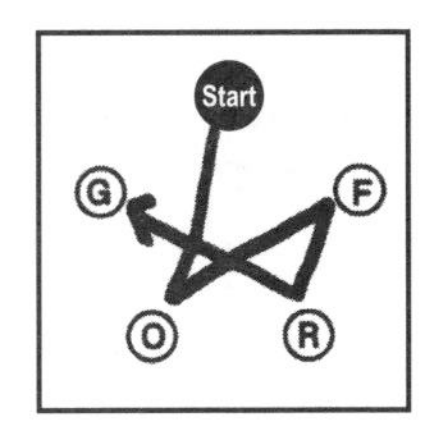

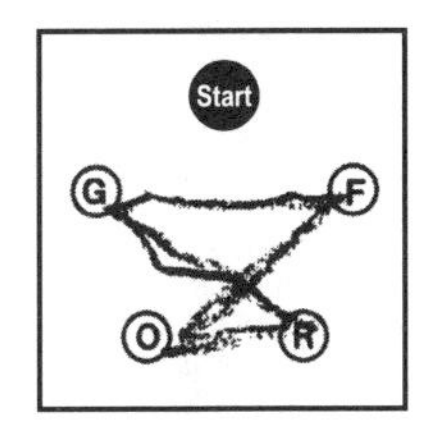

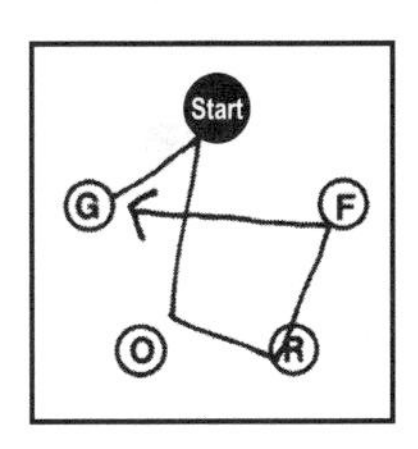

Sometimes, giving specific, concrete meaning to the process helps students see the different choices. Let **F** be Fresno, CA, **R** be Raleigh, NC, **O** be Orlando, FL, and **G** be Gary, IN. Start from another city as home, visit each of these cities just once, and then return home. Make a list of the 24 different routes you can take.

REFLECTIONS

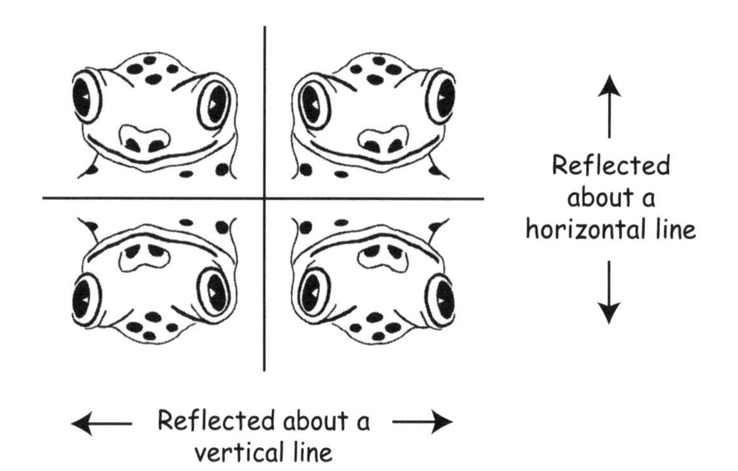

Introduction to Activity 

TRAVELING ROUTES
Testing for Traceability

Description

A key component of discrete mathematics is graph theory. This new branch of mathematics has many practical applications in the real world, but it also connects directly to many established areas of the traditional mathematics curriculum. It likewise reflects the current recommendations of national and state standards.

In Activity 6, students explore properties of networks or graphs. This leads to the discovery of a rule for tracing based on the *degrees* of the vertices. Students also determine the lengths of specific paths according to given dimensions.

This activity gives a new and refreshing approach to fundamental mathematical concepts applicable to diverse learning styles. It is rooted in graph theory, enjoyable and challenging, yet visually and intellectually stimulating.

ONE-LINE DRAWING

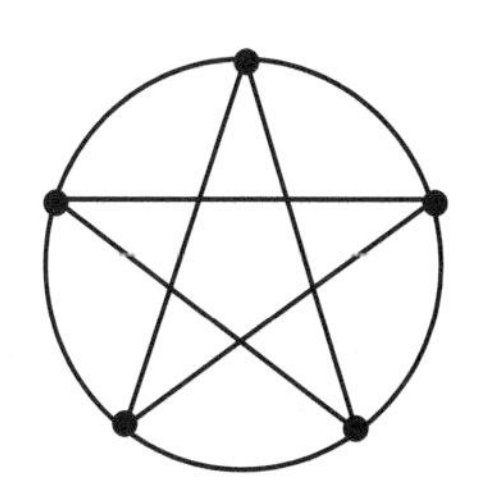

A network showing a circuit with 5 vertices and 10 edges

Big Idea

Some networks can be traced as a path by traveling over every edge exactly once. These graphs are called **one-line drawings**. If the path begins and ends at the same starting point, it is also called a *circuit*.

Expected Outcomes

Students will gain experience in

- identifying vertices and edges
- recognizing the degree of a vertex
- discovering and applying tracing rules
- locating paths and circuits
- computing and comparing lengths of paths

Meeting these NCTM Standards

✓ Numbers	✓ Problem Solving
Algebra	✓ Reasoning and Proof
✓ Geometry	✓ Communication
Measurement	✓ Connections
✓ Data Analysis	✓ Representation

Activity 6 TRAVELING ROUTES

Background

People around the world have been playing with games involving line designs for hundreds of years. The object of one such game is to try to trace the complete design without lifting the pencil off the paper and without tracing over any part of the design a second time. Designs that have this property are called ***one-line drawings***.

Vocabulary

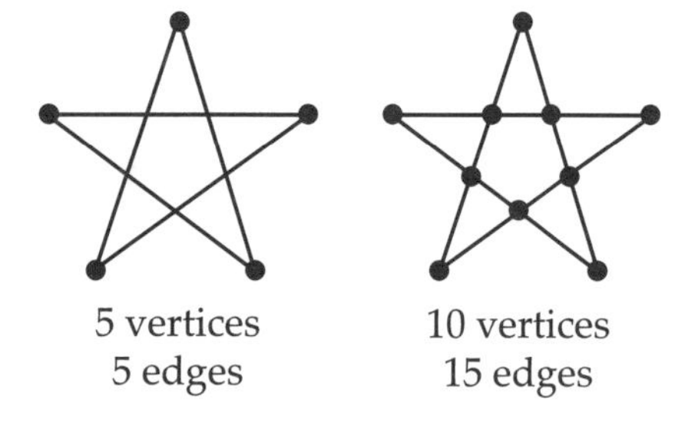

Graph theory has its own set of words to know and use when discussing one-line drawings.

vertex - a point or dot
edge - a segment or curve joining two vertices
network - a figure made up of vertices and edges
path - a route through a set of vertices
circuit - a path that begins and ends at the same vertex

First verify the given number of vertices and edges for each star design. Then, through **problem solving**, have your students show that both designs are one-line drawings. Ask if these networks show circuits or simply paths.

Activity

Try any vertex as a starting point. See if the design can be traced without lifting your pencil off the paper. You may pass through a vertex any number of times, but do not retrace any edge already traced.

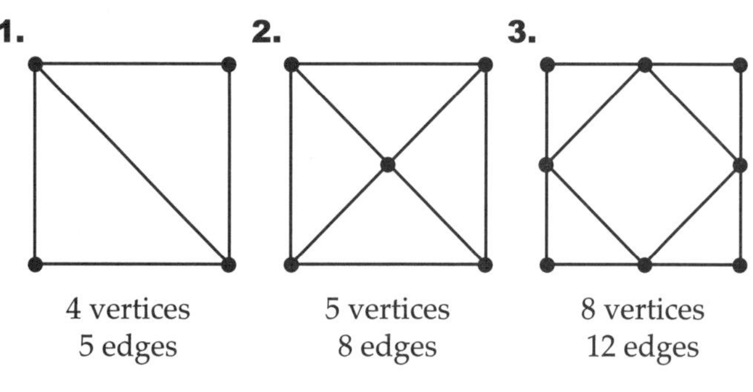

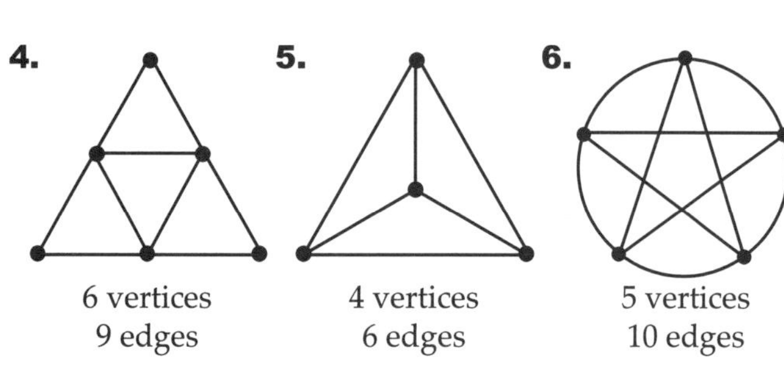

Solution

1, 3, 4, and 6 are one-line drawings.

For design 1, trace a path from the upper left, through all edges, to the lower right vertex, or the reverse.

For designs 3, 4, and 6, start at any vertex and trace a circuit through all edges, ending at that starting vertex.

The Rule for Traceability

How do you know if a design is a one-line drawing? Good **reasoning** by your students will reveal the answer. The key lies with the degrees of its vertices.

degree of a vertex - the number of edges that meet at that vertex

Circuits that traverse each edge of a graph exactly once, one-line drawings that start and end at the same point, are called ***Euler circuits***. The great Swiss mathematician, Leonhard Euler, 1707-1783, discovered this rule.

Rule: If the degrees of all vertices are even, the design can be traced as a circuit, traveling over every edge, ending back at the starting point.

If there are 2 and only 2 odd vertices, then the design can only be traced by a path starting at one odd vertex and ending at the other.

If there are more than 2 odd vertices in the design, it cannot be traced with a single line. It is not a one-line drawing.

Designs 1, 2, 3 and 4, 5, 6 are repeated here with their vertices lettered. The table below lists the vertices for each network as even or odd and tells which of the designs are **one-line drawings**.

Have your students create a similar table for these designs. You can find two transparency masters for these same figures on pages 74 and 75.

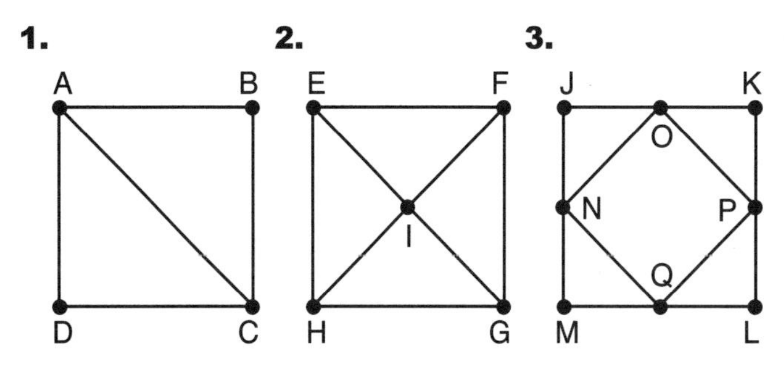

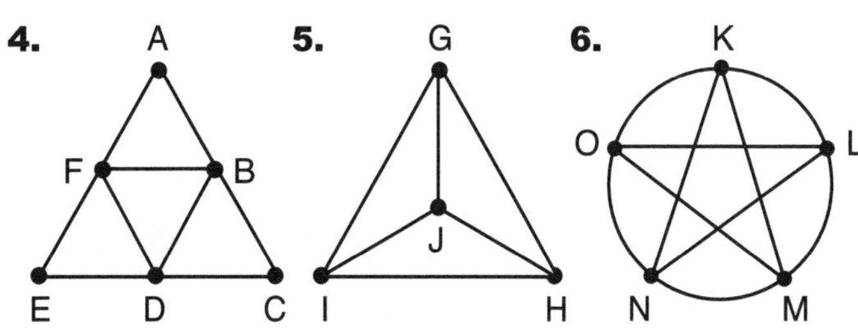

Design	Even vertices	Odd vertices	One-line drawing	Starting vertex	Path or circuit
1	B,D	A,C	yes	A or C	path
2	I	E,F,G,H	no	none	neither
3	J,K,L,M,N,O,P,Q	none	yes	any vertex	circuit
4	A,B,C,D,E,F	none	yes	any vertex	circuit
5	G,H,I	J	no	none	neither
6	K,L,M,N,O	none	yes	any vertex	circuit

ONE-LINE DRAWINGS

Worksheet 6A

In a one-line drawing, the entire design can be traced
without lifting the pencil off the paper, and
without tracing over any edge twice.
Any vertex can be passed through any number of times.

- Count the number of edges coming from the vertices, shown as circles. Enter those numbers in the circles.
- Find the number of odd vertices in each network.
- Is the network a one-line drawing? Answer **yes** or **no**. If the answer is yes, show how it can be traced.

1.

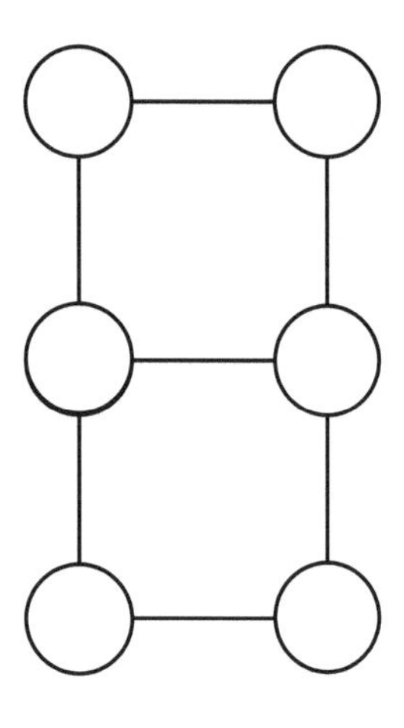

2.

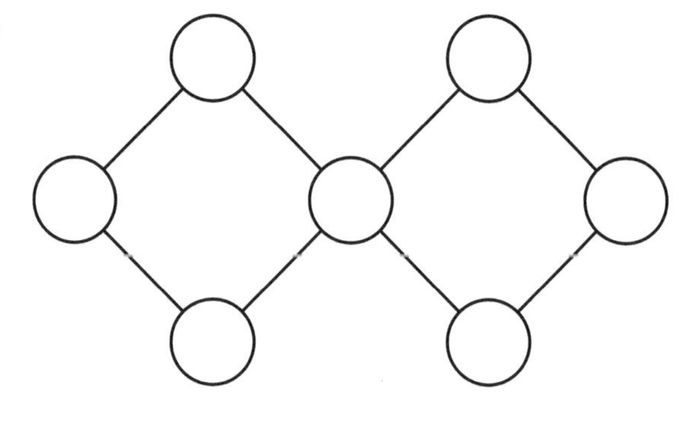

3.

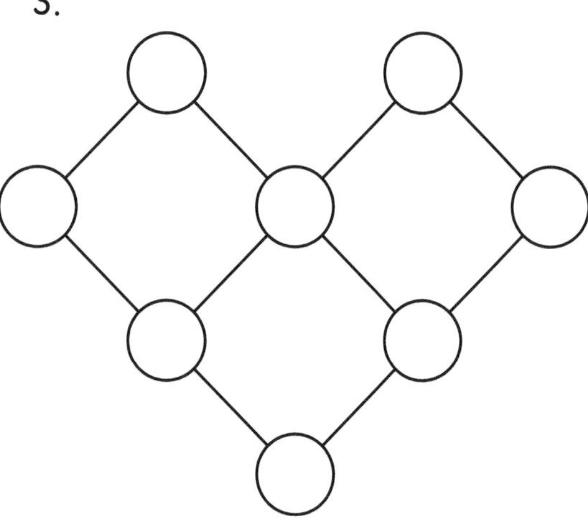

4.

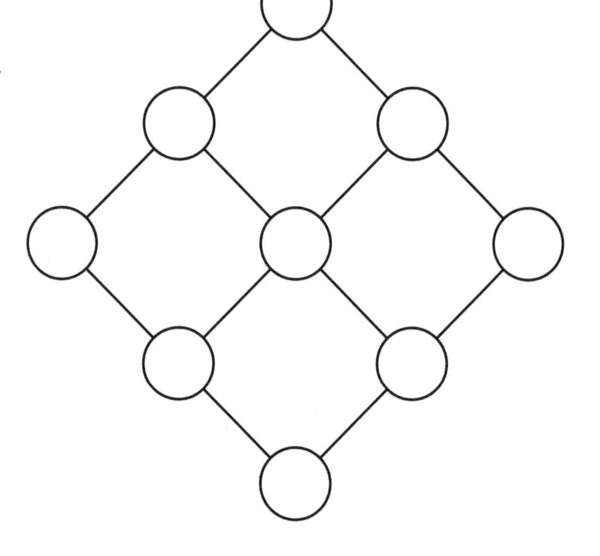

SHORTEST PATH

Worksheet 6B

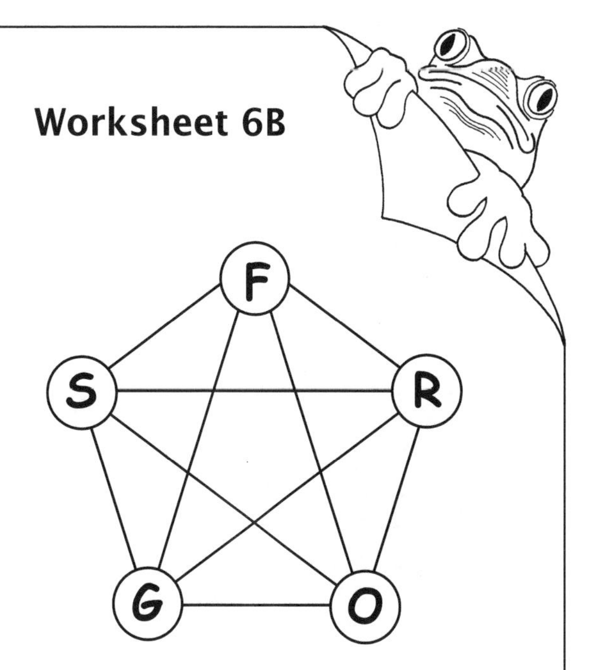

This network has the five letters in the word FROGS for its vertices.

1. How many vertices are in the network?

2. How many edges are in the network?

3. Can you trace over the entire design without lifting your pencil off the paper and without tracing over any edge twice?

4. Is the design a one-line drawing?

To assign a dimension to an edge, find the numbers for its 2 lettered endpoints. Subtract them and use the difference as the dimension.

A	B	C	D	E	F	G	H	I	J	K	L	M	N	O	P	Q	R	S	T	U	V	W	X	Y	Z
1	2	3	4	5	6	7	8	9	10	11	12	13	14	15	16	17	18	19	20	21	22	23	24	25	26

For example,

edge OR $18 - 15 = 3$
edge FO $15 - 6 = 9$

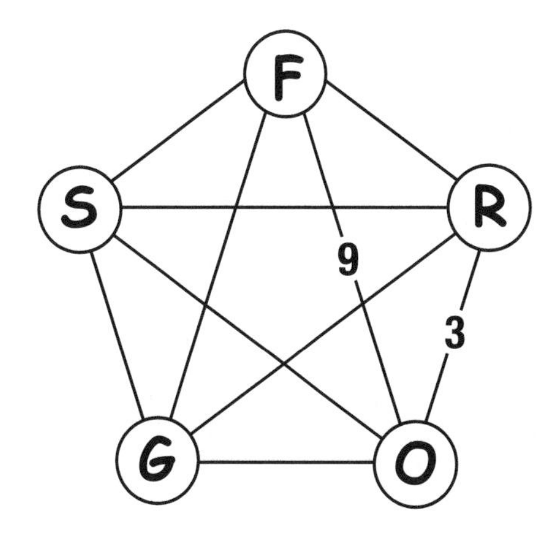

5. Find the lengths of the 8 remaining edges and mark them on the network.

6. Show that the path, FROGS, has a length of 35.

7. Find the length of the path GOFRS.

8. Find the length of the path FOSRG.

9. Find and name the shortest path from F that connects the remaining 4 vertices. Give its length and then trace it out on the network.

ANSWERS to Worksheets 6A and 6B

ONE-LINE DRAWINGS

Worksheet 6A

In a one-line drawing, the entire design can be traced
without lifting the pencil off the paper, and
without tracing over any edge twice.
Any vertex can be passed through any number of times.

- Count the number of edges coming from the vertices, shown as circles. Enter those numbers in the circles.
- Find the number of odd vertices in each graph.
- Is the graph a one-line drawing? Answer **yes** or **no**. If the answer is yes, show how it can be traced.

1. **Yes. Start at either of the two odd vertices.**
2. **Yes. Start at any vertex.**
3. **Yes. Start at either of the two odd vertices.**
4. **No. There are more than two odd vertices.**

70

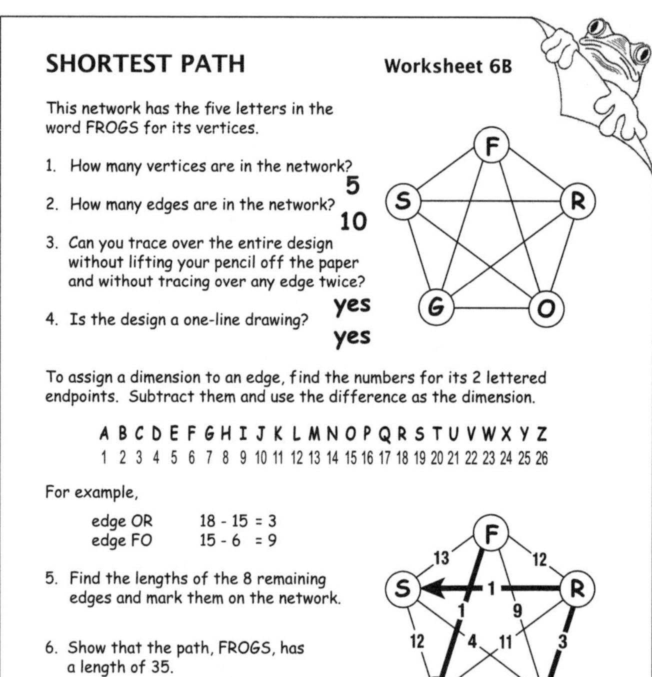

SHORTEST PATH

Worksheet 6B

This network has the five letters in the word FROGS for its vertices.

1. How many vertices are in the network? **5**
2. How many edges are in the network? **10**
3. Can you trace over the entire design without lifting your pencil off the paper and without tracing over any edge twice? **yes**
4. Is the design a one-line drawing? **yes**

To assign a dimension to an edge, find the numbers for its 2 lettered endpoints. Subtract them and use the difference as the dimension.

A	B	C	D	E	F	G	H	I	J	K	L	M	N	O	P	Q	R	S	T	U	V	W	X	Y	Z
1	2	3	4	5	6	7	8	9	10	11	12	13	14	15	16	17	18	19	20	21	22	23	24	25	26

For example,

edge OR 18 - 15 = 3
edge FO 15 - 6 = 9

5. Find the lengths of the 8 remaining edges and mark them on the network.
6. Show that the path, FROGS, has a length of 35.
   12+3+8+12 = 35
7. Find the length of the path GOFRS.
   8+9+12+1 = 30
8. Find the length of the path FOSRG.
   9+4+1+11 = 25
9. Find and name the shortest path from F that connects the remaining 4 vertices. Give its length and then trace it out on the network.
   FGORS 1+8+3+1 = 13

71

FROG Facts

The bulging eyes of a frog are binocular in that each eye can see in a different direction at the same time. Frogs tend to be farsighted and will often miss food that is right under their noses unless it moves. They have a second pair of transparent eyelids that they use to protect their eyes when jumping and swimming.

TRANSPARENCY MASTERS

Make transparencies from these masters.

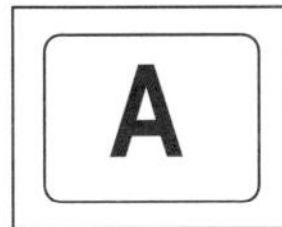

Use this transparency in conjunction with designs 1, 2, and 3 shown on page 69.

Use this transparency in conjunction with designs 4, 5, and 6 shown on page 69.

Discuss all six designs shown on transparencies A and B with your students before using the student worksheets.

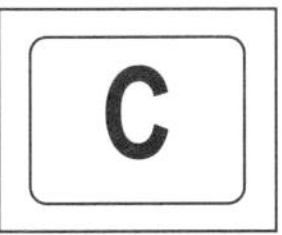

This transparency may be helpful when used with Worksheet 6B. It contains the dimensions of all 10 edges in the FROGS network shown there. Use it as an answer key for students to check their work.

On this transparency, you can create your own dimensions for the 10 edges of the same network. Use any word with 5 different letters to identify the 5 vertices. Then let students compute the dimensions of the 10 edges by subtraction using the alphabet code supplied. Once complete, have them use the network to find the lengths of some given paths, and then let them search for the shortest path that connects all five vertices. The shortest path will be the one that takes the letters either in alphabetic order or in reverse alphabetic order.

ONE-LINE DRAWINGS

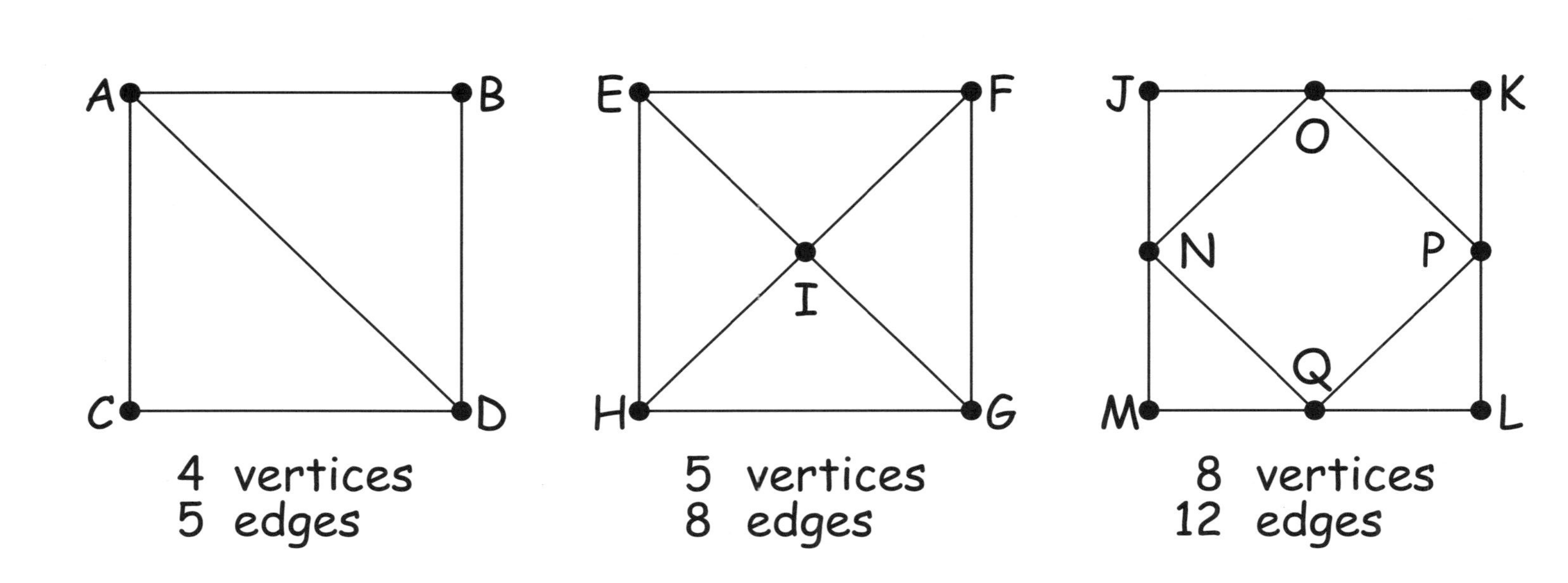

Try any vertex as the starting point. See if you can trace the design without lifting your pencil off the paper. Do not retrace any edge already traced.

A

ONE-LINE DRAWINGS

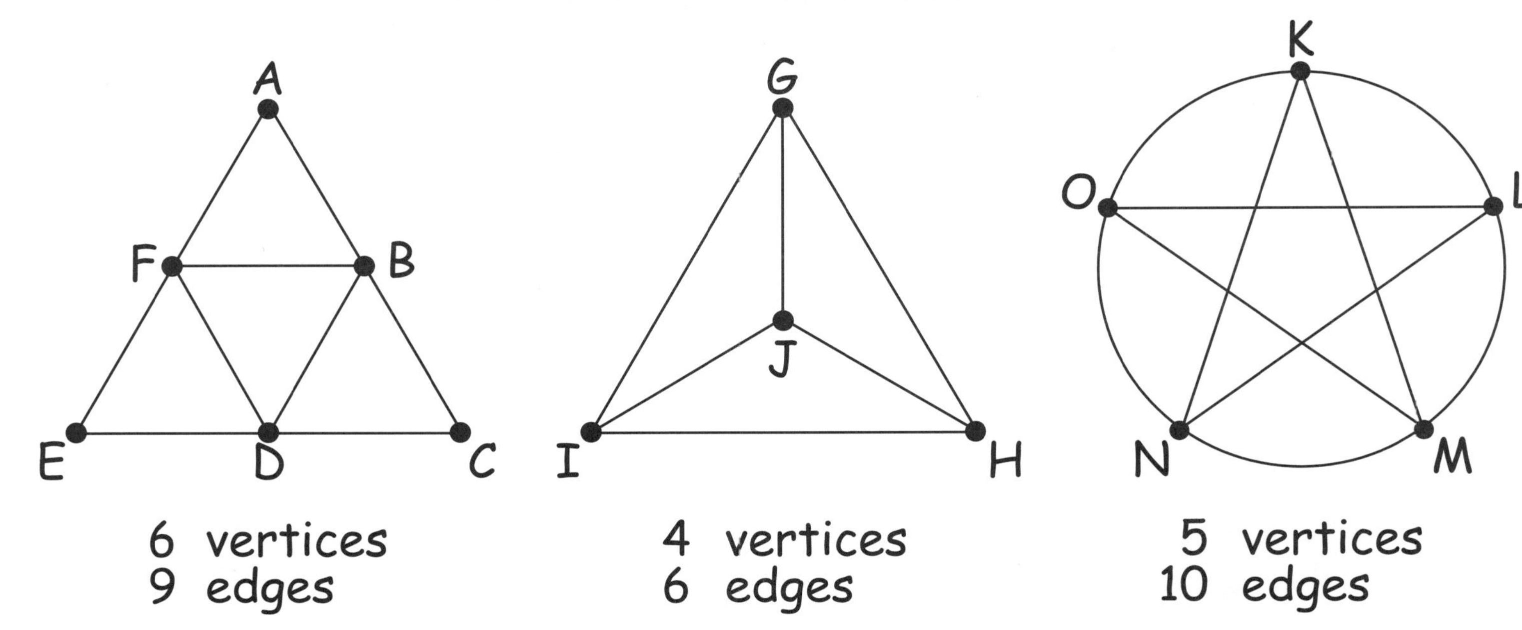

Try any vertex as the starting point. See if you can trace the design without lifting your pencil off the paper. Do not retrace any edge already traced.

B

SHORTEST PATH

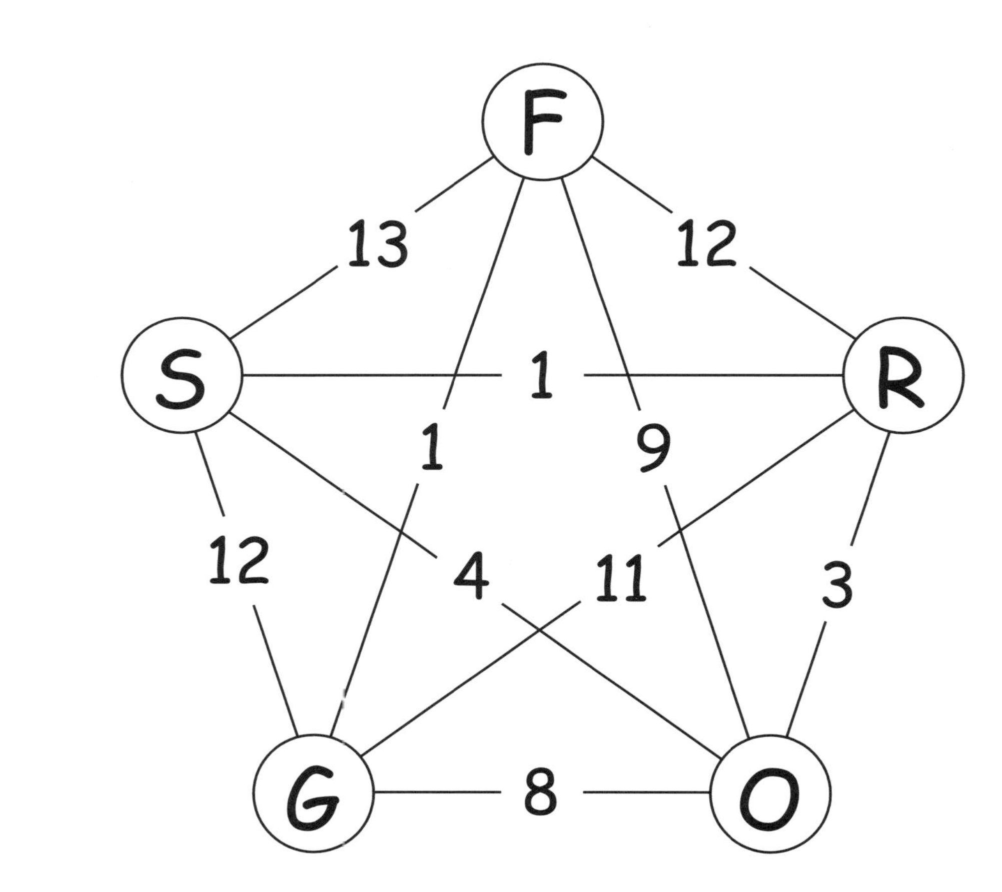

Find the shortest path from vertex F that connects the remaining 4 vertices.

C

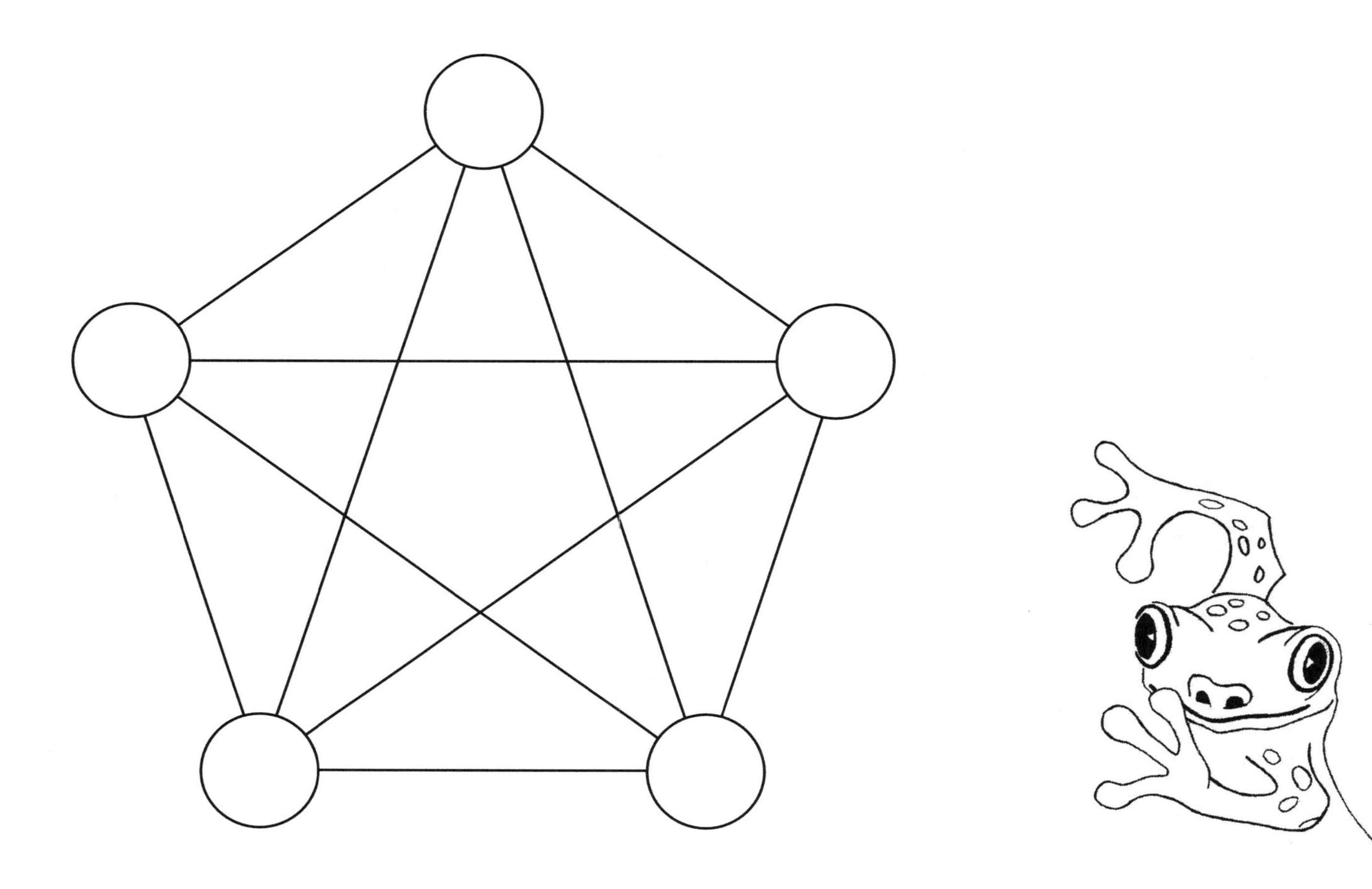

A	B	C	D	E	F	G	H	I	J	K	L	M	N	O	P	Q	R	S	T	U	V	W	X	Y	Z
1	2	3	4	5	6	7	8	9	10	11	12	13	14	15	16	17	18	19	20	21	22	23	24	25	26

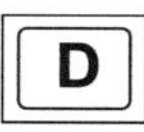

GETTING THE MOST from Activity 6

This activity uses **one-line drawings** as a simple, enjoyable way to help students learn the vocabulary related to graph theory. Through it they see a need for identifying and labeling vertices on a network. They also have an opportunity to contrast the term *edges* in graph theory with *edges* in geometric space figures. Use the two transparencies, A and B, with this activity.

vertex
edge
network
path
circuit

Students can follow the designs on the two transparencies or be supplied with copies of their own as you guide the discussion. You can let your students predict which designs on each page can be traced. See if they can explain the **reasoning** they use.

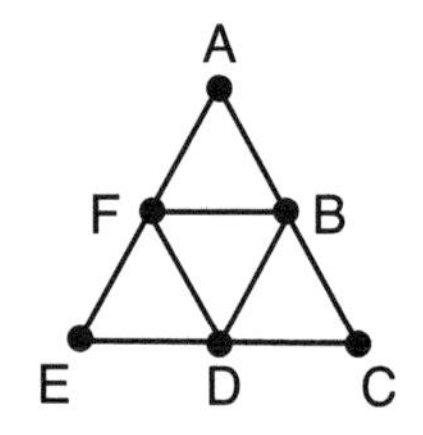

To facilitate the discussion and understanding of which vertices are being used, the vertices are labeled. This letter **representation** will help in describing paths and in focusing on the specific starting positions.

Any of the six lettered vertices, A, B, C, D, E, and F, can be used as the starting point in the first graph shown, but only vertices P and R can be used as starting points in the second graph.

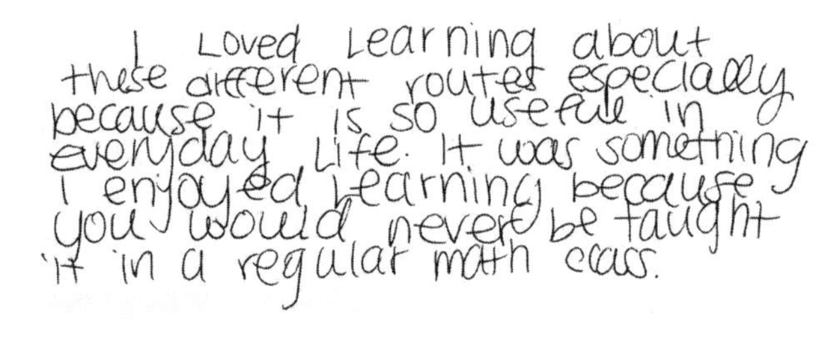

Here is a student that recognized both the usefulness and uniqueness of this topic.

Good **problem-solving** situations often contain added features that students can discover and explore on their own. We learned that circuits that traverse each edge of a graph exactly once, starting and ending at the same point, are called **Euler circuits.** These are one-line drawings. Encourage your students to investigate another kind of circuit that require traveling through each vertex exactly once. These are called **Hamiltonian circuits.**

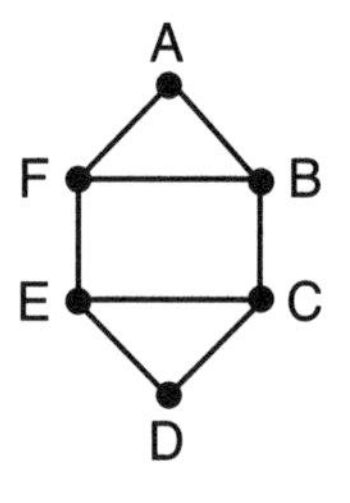

The first network shown at the right contains a Hamiltonian circuit, but it is not a one-line drawing. The second network does not contain a Hamiltonian circuit, but it is a one-line drawing.

STUDENT REACTIONS to Activity 6

These student solutions focus on the details of the activity.

In this first solution,

- the student indicates that the design can be traced by answering *yes*.
- the vertices have been labeled to facilitate recording the sequence that was followed.
- the starting and ending points are clearly noted on the drawing.
- the route is identified as a path but not a circuit since the starting and ending points are not the same.

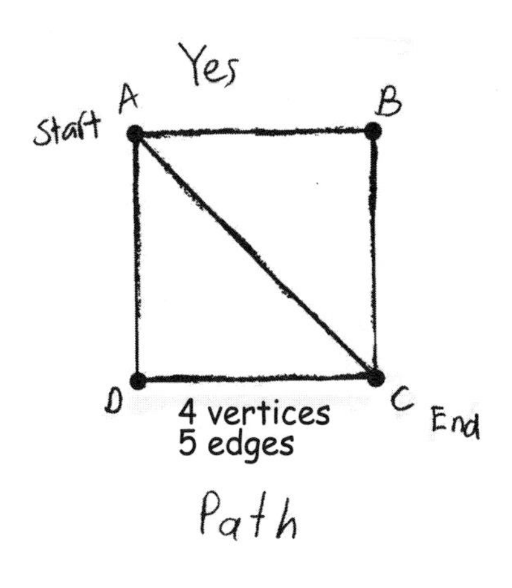

The path chosen by the student was ABCADC. In all, six different paths from A are possible, but they all end at C.

ABCADC	ACBADC	ADCABC
ABCDAC	ACDABC	ADCBAC

Students will be surprised to find that this second design cannot be traced without going over at least one edge twice. It is not a one-line drawing. However, you can travel through every vertex of this same graph exactly once without going through any vertex twice.

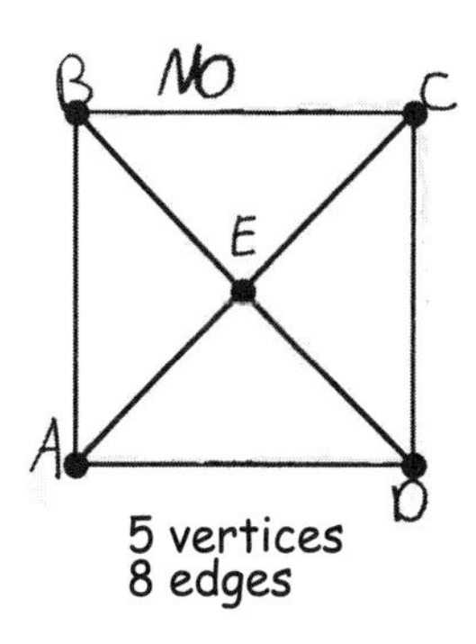

In the third example, the student again finds that the graph can be traced. It is another one-line drawing. However, this time it is possible to start and end at the same vertex so it is more than just a path. It is a circuit, an Euler circuit.

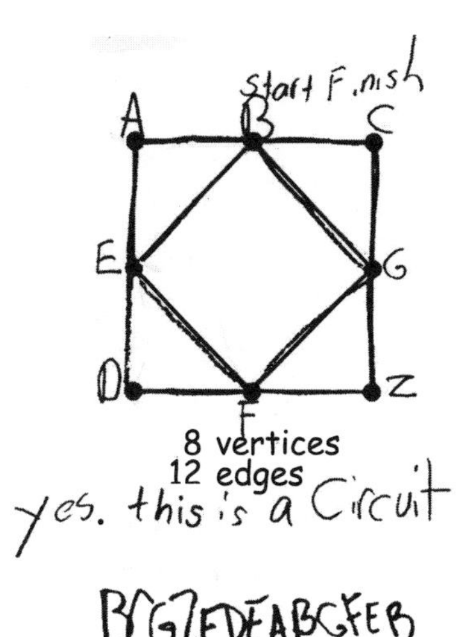

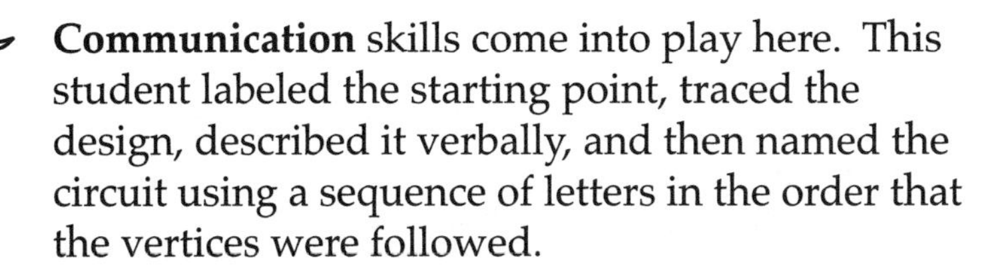

Communication skills come into play here. This student labeled the starting point, traced the design, described it verbally, and then named the circuit using a sequence of letters in the order that the vertices were followed.

RATIOS

The familiar Spring Peeper is about 1 inch long.
The giant African goliath frog is about 15 inches long.
The ratio of body lengths is 1 to 15. The ratio of body volumes is 1 to 3375.

Introduction to Activity 

REPEATING STEPS
Drawing Repeating Patterns

Description

Many discrete math activities involve iteration, where a process is repeated over and over again. Activity 7 creates geometric designs from words coded into repeating digit sequences. These repeating digit sequences are then used as lengths of successive line segments drawn on a square dot grid. The resulting iterative designs are called *spirolaterals*.

CODE			
1	A	J	S
2	B	K	T
3	C	L	U
4	D	M	V
5	E	N	W
6	F	O	X
7	G	P	Y
8	H	Q	Z
9	I	R	

For the word **TONGUE** (**265735**), move **2** steps right from the marked starting point, turn and move **6** steps down. Then move **5** steps left, **7** up, **3** right, and **5** down. Two complete cycles through the digits, turning as you go, bring you back to the starting point.

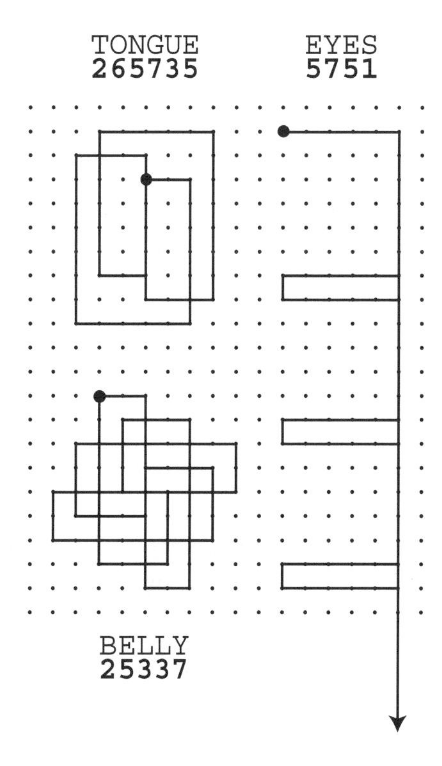

Big Idea

Two repeating patterns were used in creating these designs. One repeats numbers while the other repeats turns. **Iteration** is the act of repeating steps over and over. Drawing a spirolateral is an iterative process.

Expected Outcomes

Students will gain experience in

- following directions
- repeating steps
- counting carefully
- seeing geometric iteration patterns
- recognizing rotational symmetry

Meeting these NCTM Standards

✓ Numbers
Algebra
✓ Geometry
✓ Measurement
Data Analysis
✓ Problem Solving
Reasoning and Proof
Communication
✓ Connections
✓ Representation

REPEATING STEPS

Getting Ready

CODE			
1	A	J	S
2	B	K	T
3	C	L	U
4	D	M	V
5	E	N	W
6	F	O	X
7	G	P	Y
8	H	Q	Z
9	I	R	

Use a sheet of square dot paper.

Mark a dot as a starting point.

With the letter-digit code, translate the word **FROGS** to the number **69671**.

Think of the number **69671** as a sequence of digits, **6**, **9**, **6**, **7**, and **1**.

Creating the Figure

From the starting point, move **6** steps right, **9** down, **6** left, **7** up, and then **1** right.

Repeat the digit sequence, over and over.

6,9,6,7,1,6,9,6,7,1,6,9,6,7,1,6,9,6,7,1,...

At the same time, keep turning clockwise, always repeating the same direction steps, Right, Down, Left, Up, over and over.

R,D,L,U,R,D,L,U,R,D,L,U,R,D,L,U,R,D,L,U,...

Watch for something special to happen.

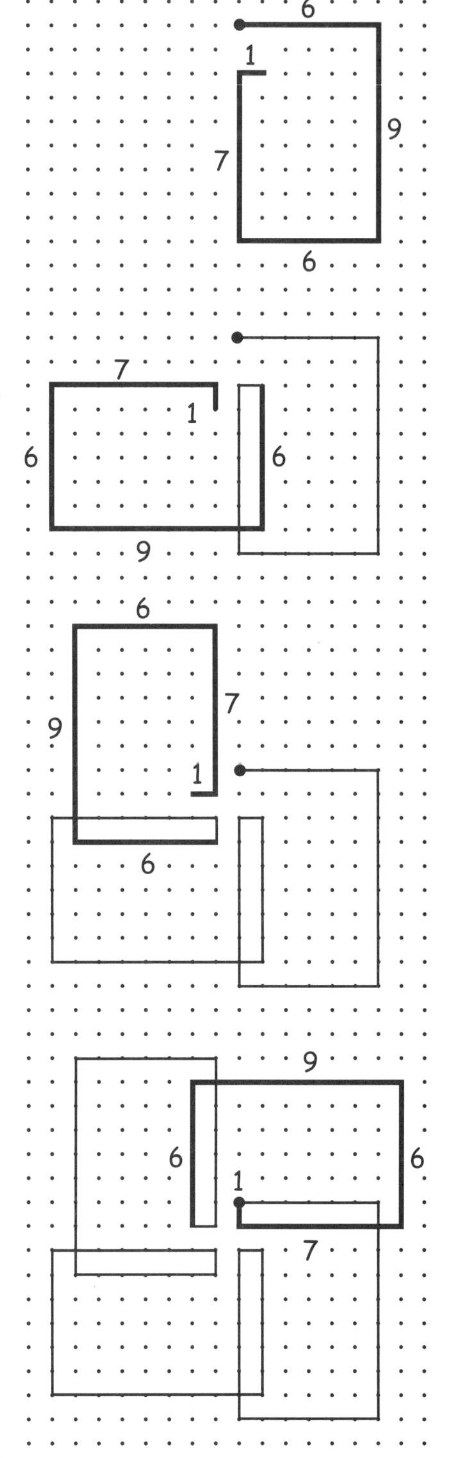

Studying the Result

After 4 repeated cycles through this **iteration** using the word **FROGS**, coded as the repeating digit sequence **6**, **9**, **6**, **7**, **1**, the path returns to the starting point. Thereafter, additional steps trace over the existing path.

For students who have trouble following the digit and direction sequences at the same time, consider creating this table first. It will help them match digits to directions.

6,9,6,7,1,6,9,6,7,1,6,9,6,7,1,6,9,6,7,1,...
R,D,L,U,R,D,L,U,R,D,L,U,R,D,L,U,R,D,L,U,...

Properties

Figures of this kind are called spirolaterals. Whatever the original starting word used in the activity, only three different types of spirolaterals are possible.

A word such as **FROGS** produces a spirolateral that retraces itself after 4 cycles.
A word such as **LIZARD** produces one that retraces itself after 2 cycles.
A word such as **TOAD** generates a spirolateral that never ever comes back to its starting point but simply moves off the page.

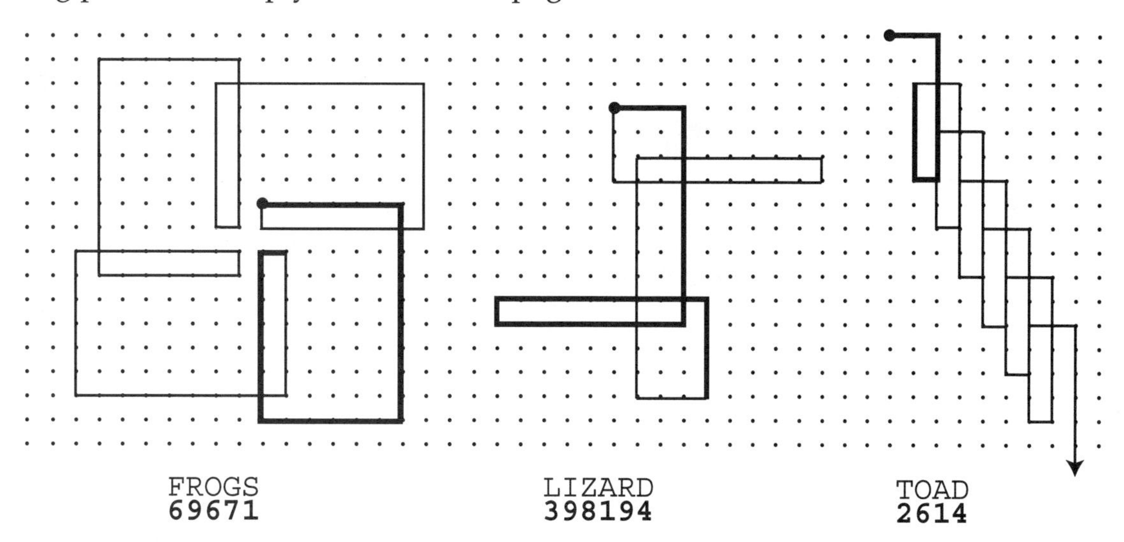

Patterns

The number of letters determines the particular type of spirolateral produced.

Number of letters	2	3	4	5	6	7	8	9	10	11	12
Number of cycles	2	4	Spin off	4	2	4	Spin off	4	2	4	Spin off

Extension

Try having your students make spirolaterals as **representations** for their names. All students follow the same procedure, but each one creates a different design.

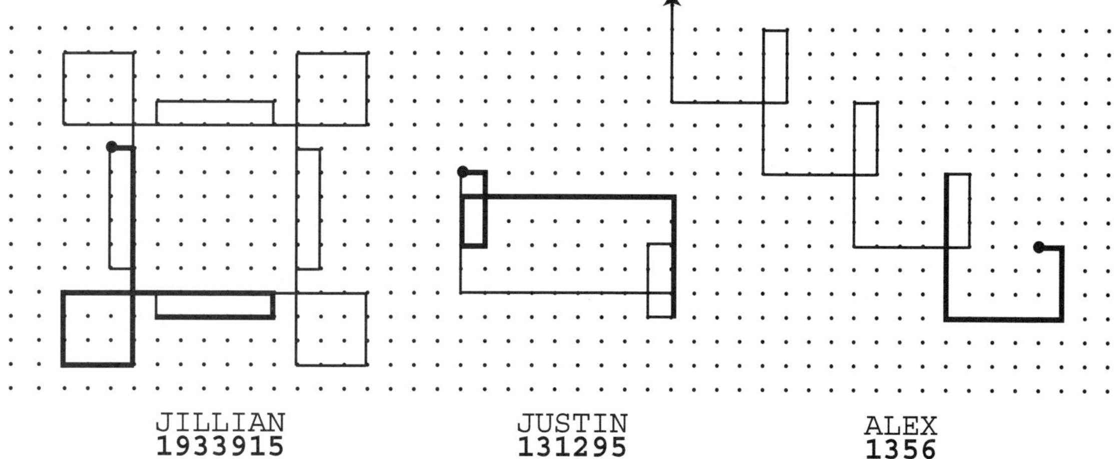

SPIROLATERALS

Worksheet 7A

Try your hand at drawing spirolaterals for the words TURTLE and TADPOLE.

TURTLE
239235

The first six segments are drawn. Trace them out.

2 right
3 down
9 left
2 up
3 right
5 down

Now draw the next six segments to repeat the pattern and complete the spirolateral.

2 left
3 up
9 right
2 down
3 left
5 up

TADPOLE
2147635

Start at the marked point. Draw 2 right, 1 down, 4 left, 7 up, 6 right, 3 down, and 5 left. Repeat the sequence of digits three more times. Continue the clockwise turns as you go. You should end at the starting point.

YOUR SPIROLATERAL

Worksheet 7B

1	A J S
2	B K T
3	C L U
4	D M V
5	E N W
6	F O X
7	G P Y
8	H Q Z
9	I R

Print the letters in your first name.
Code them into a sequence of digits.

___ ___ ___ ___ ___ ___ ___ ___ ___ ___

___ ___ ___ ___ ___ ___ ___ ___ ___ ___

Mark a starting point. Follow your sequence of digits. Keep moving clockwise, right, down, left, and up. Repeat until the design is complete or spins off.

ANSWERS to Worksheets 7A and 7B

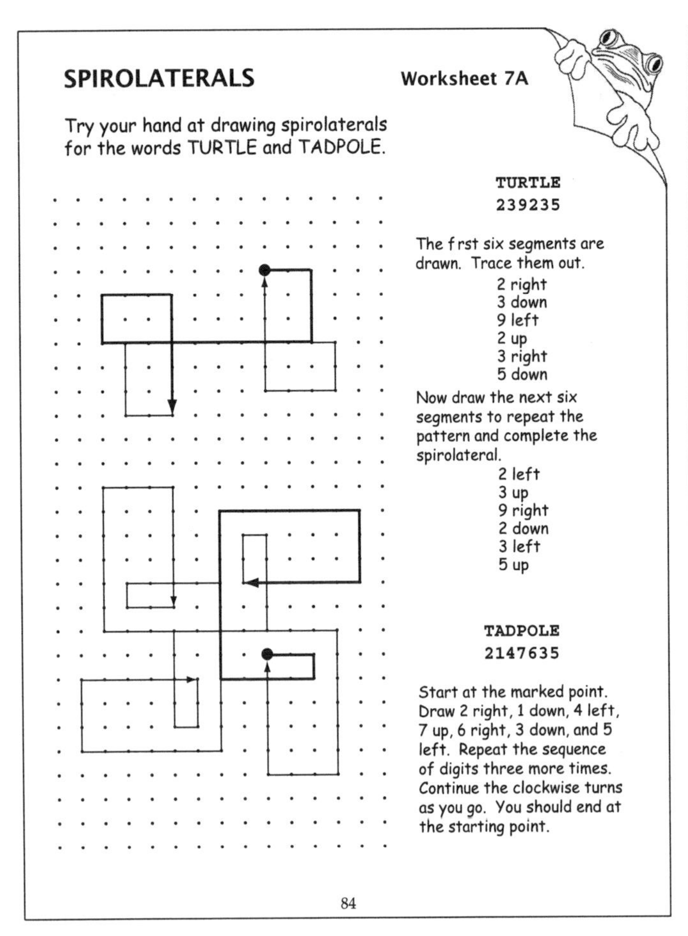

SPIROLATERALS Worksheet 7A

Try your hand at drawing spirolaterals for the words TURTLE and TADPOLE.

TURTLE
239235

The first six segments are drawn. Trace them out.
2 right
3 down
9 left
2 up
3 right
5 down

Now draw the next six segments to repeat the pattern and complete the spirolateral.
2 left
3 up
9 right
2 down
3 left
5 up

TADPOLE
2147635

Start at the marked point. Draw 2 right, 1 down, 4 left, 7 up, 6 right, 3 down, and 5 left. Repeat the sequence of digits three more times. Continue the clockwise turns as you go. You should end at the starting point.

84

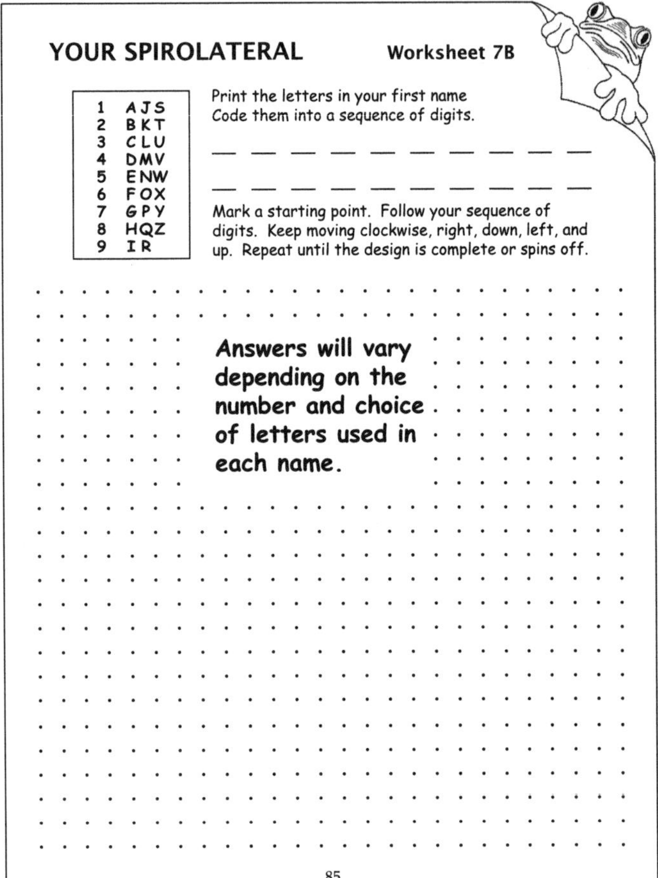

YOUR SPIROLATERAL Worksheet 7B

1	AJS
2	BKT
3	CLU
4	DMV
5	ENW
6	FOX
7	GPY
8	HQZ
9	IR

Print the letters in your first name
Code them into a sequence of digits.

__ __ __ __ __ __ __ __ __ __

__ __ __ __ __ __ __ __ __ __

Mark a starting point. Follow your sequence of digits. Keep moving clockwise, right, down, left, and up. Repeat until the design is complete or spins off.

Answers will vary depending on the number and choice of letters used in each name.

85

FROG Facts

Each frog has its own special call that it repeats over and over. These may be croaks, clucks, grunts, snores, growls, chirps, trills, whistles, or other strange sounds. Try repeating these frog calls.

Pe-ep, pe-ep, pe-ep, pe-ep, pe-ep
Ribid, ribid, ribid, ribid, ribid
Jug-o-rum, jug-o-rum, jug-o-rum, jug-o-rum
Arrk, arrk, arrk, arrk, arrk, arrk

TRANSPARENCY MASTERS

Make transparencies from these masters.

Use this transparency to discuss the letter-digit code. Show how the 5-letter sequence in the word FROGS translates into the 5-digit sequence 69671 and then into the corresponding spirolateral. When projected on the screen, start at the marked point and trace out the path with your finger. Move 6 steps right, 9 down, 6 left, 7 up, and 1 right. Then show how this same number sequence repeats itself three more times before the figure is complete.

This transparency will help students see how the 5-digit cycle is repeated 4 times in the spirolateral. It also shows how the 5-digit number cycle, 69671, is matched against the 4-position direction cycle, RDLU. Point out that the initial 6 in the number sequence is matched first with an R, then with a D, an L, and finally a U before repeating.

Students will discover on their own the three different types of designs that can be created on square dot paper by this process. This transparency shows each type using only digit sequences.

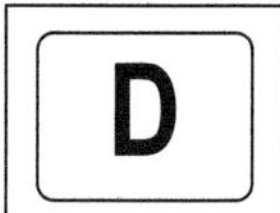

Use this square dot grid to help students draw spirolaterals. Have students draw their own individual spirolaterals from the coded number sequences created from the letters in their first names. Note that the choice of location for the starting point may be critical when trying to fit the figure on the grid.

SPROLATERAL Letter–Digit Code

1	A	J	S
2	B	K	T
3	C	L	U
4	D	M	V
5	E	N	W
6	F	O	X
7	G	P	Y
8	H	Q	Z
9	I	R	

FROGS
69671

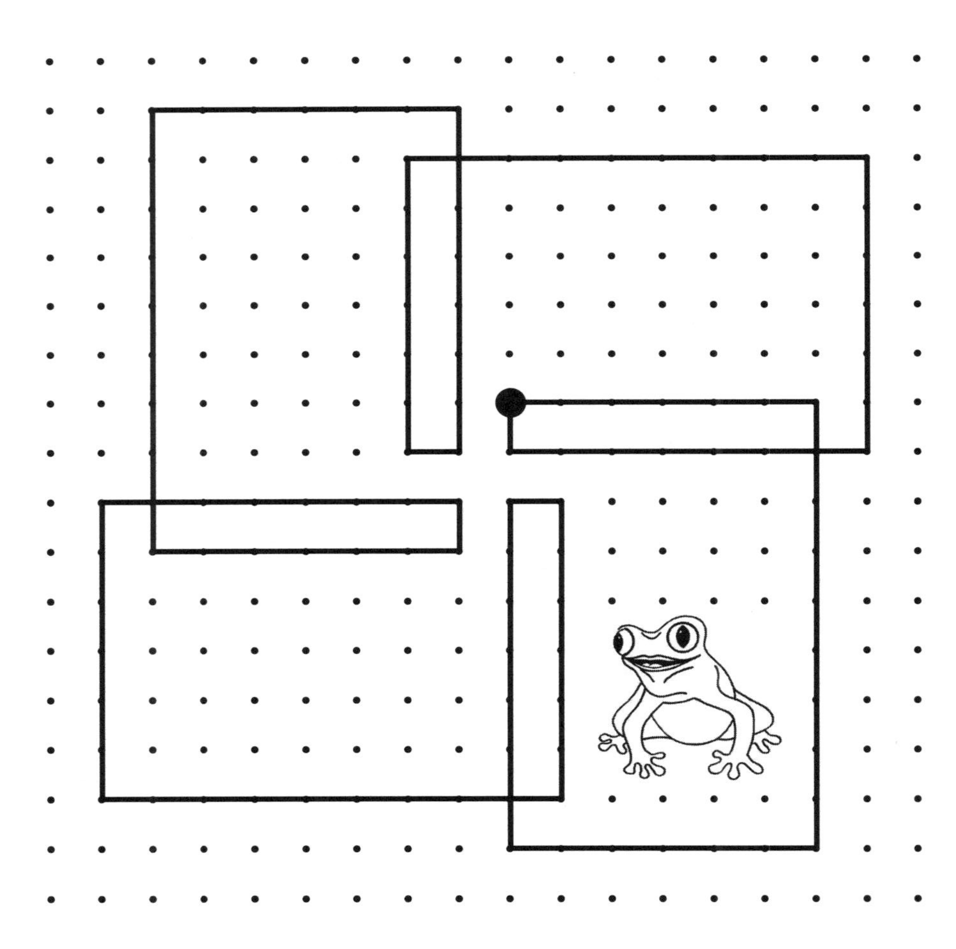

A

SPIROLATERAL
for the word FROGS
coded as 69671

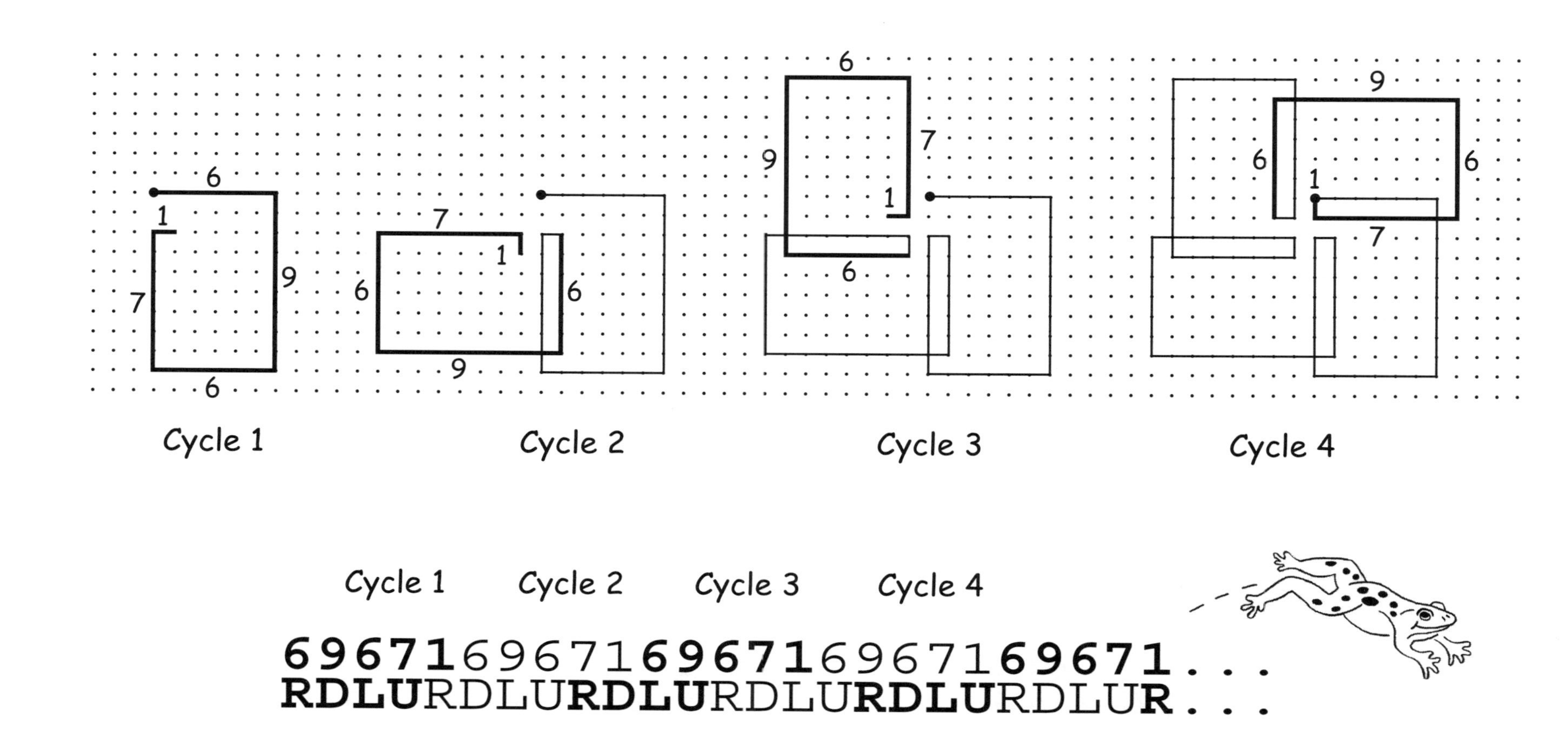

B

SPIROLATERALS by the Numbers

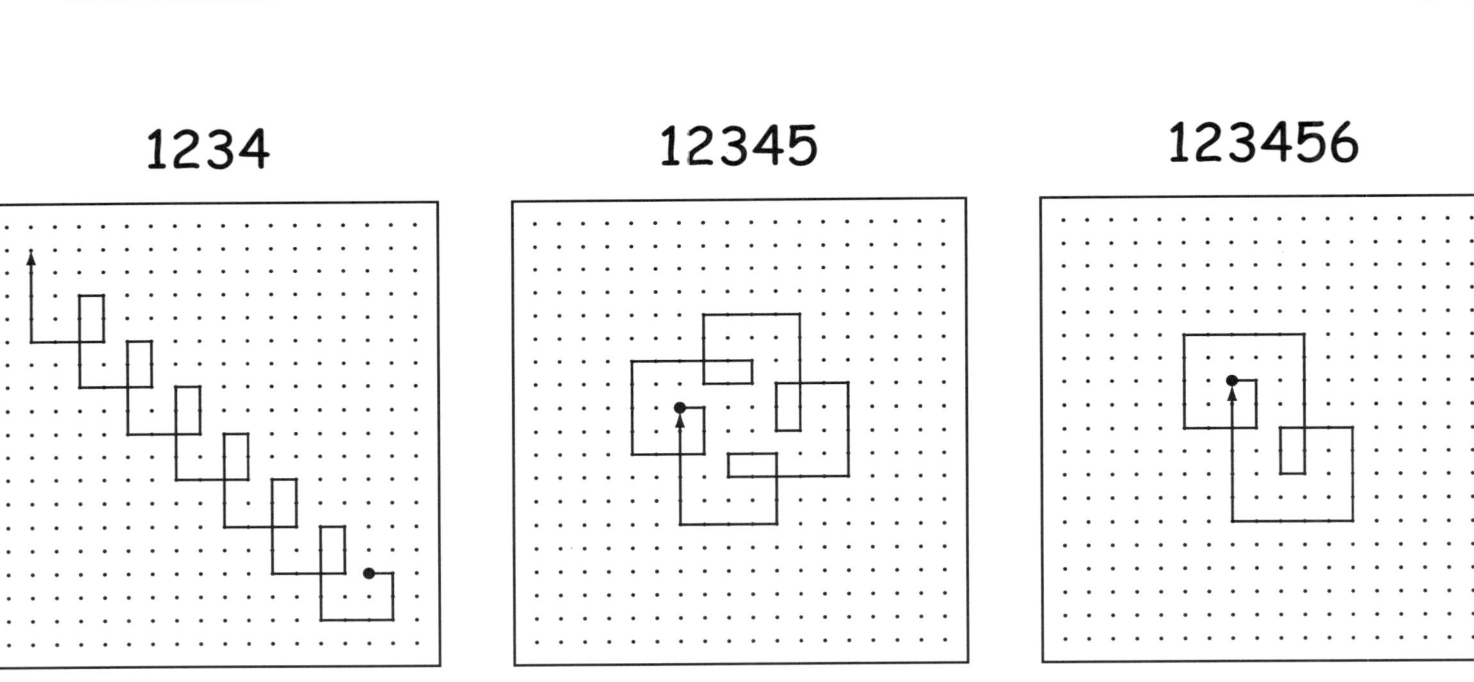

C

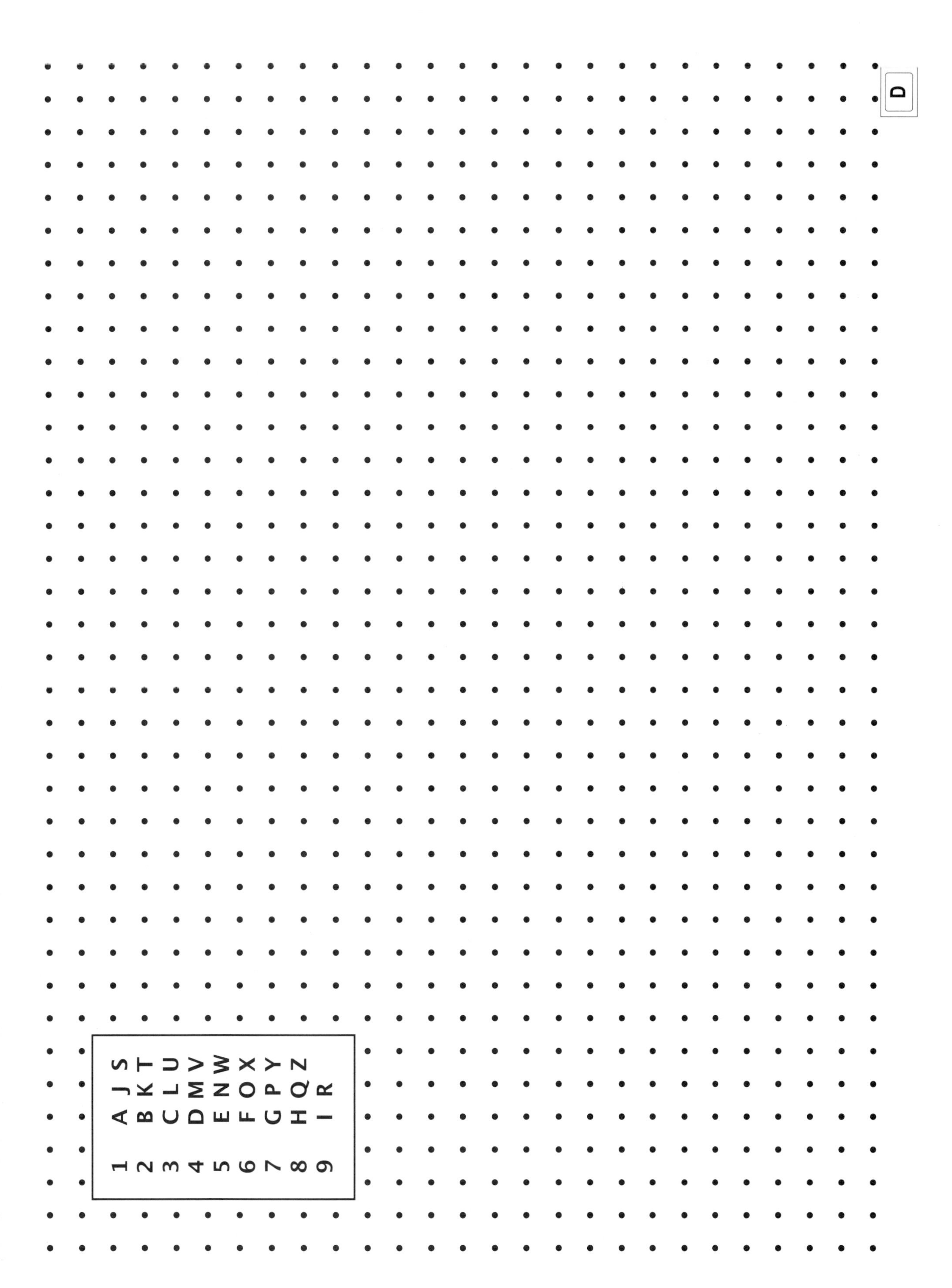
D
1 A J S
2 B K T
3 C L U
4 D M V
5 E N W
6 F O X
7 G P Y
8 H Q Z
9 I R

GETTING THE MOST from Activity 7

Be sure students use the square dot grid correctly.

- Watch for students who count dots instead of counting steps between dots.
- Remember, lines can cross each other, and they can overlap each other.
- Always repeat the full number sequence each time, even when the first and last digits are the same.

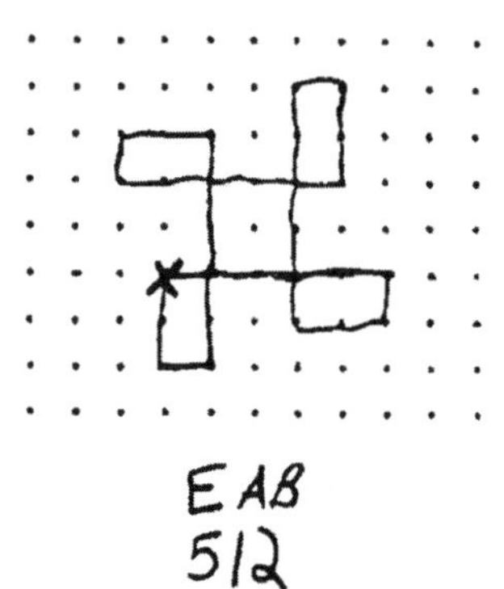

Two **iteration** sequences are being used simultaneously when drawing a spirolateral. This may be difficult for some students to follow, so try something simple first.

- Consider having all students begin with their three initials. This way, all their spirolaterals will have the same basic 4-part characteristics as shown here.

This activity shows a **connection** between numbers and pictures and also between digits and directions. Students may have trouble dealing with both issues at the same time.

The hardest part was going in the right direction.

Offer help to those students who may need it by suggesting they first write out a string of repeating digits and then match each digit to one of the four directions, using R, D, L, and U.

Here are the two matched sequences for the set of initials shown above. Notice that after 4 complete cycles, the matching begins to repeat again.

EAB

512512512512512512...
RDLURDLURDLURDLURD...

Let students do their own **problem solving** in this activity by reversing the process. Show them a completed spirolateral and see if they can discover the word that was used to create it. You might code your own name or perhaps use the name of some famous mathematician from the past.

Here is how one student drew the spirolateral for the name of a famous French mathematician, Pascal (1623-1662). He then showed it to his classmates and challenged them to unravel the code and identify the mathematician. The digit-code for PASCAL is 711313.

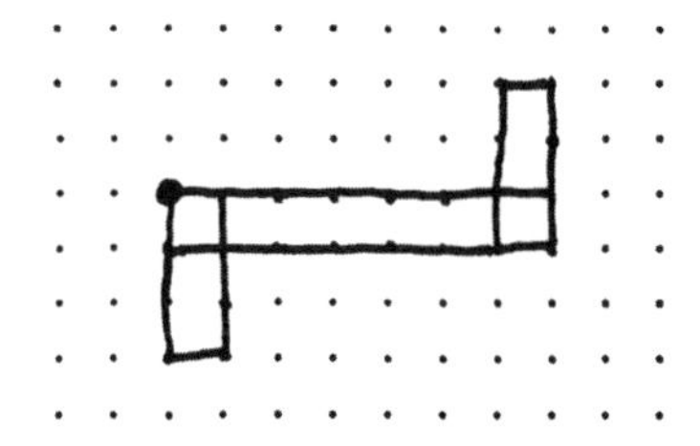

STUDENT REACTIONS to Activity 7

This is how two students, Jillian and William, drew their spirolaterals. Even though the middle five numbers in the codes for their 7-letter first names are the same, the two designs are quite different. Figures have been reduced in size.

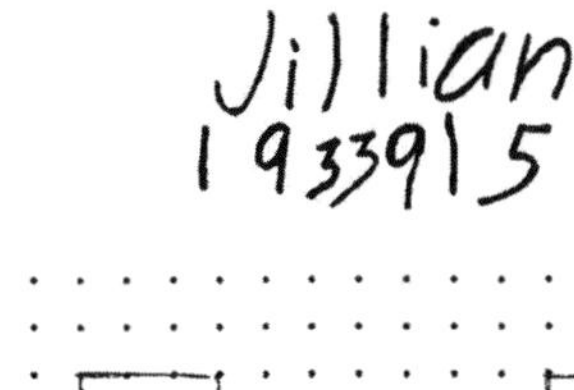

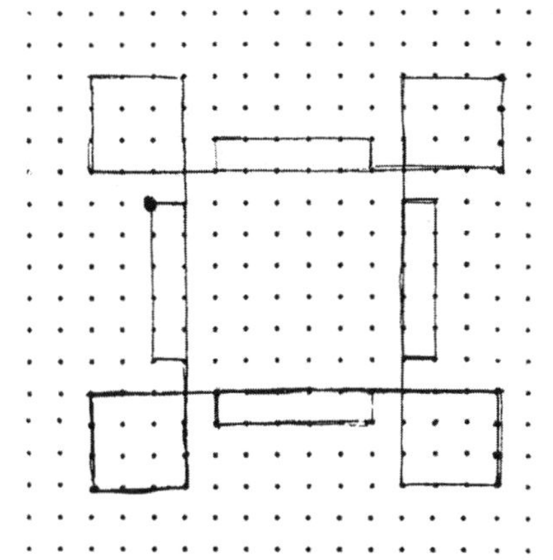

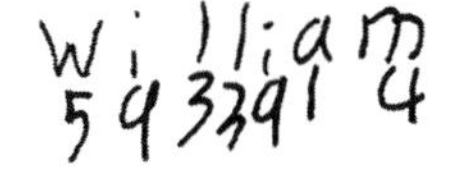

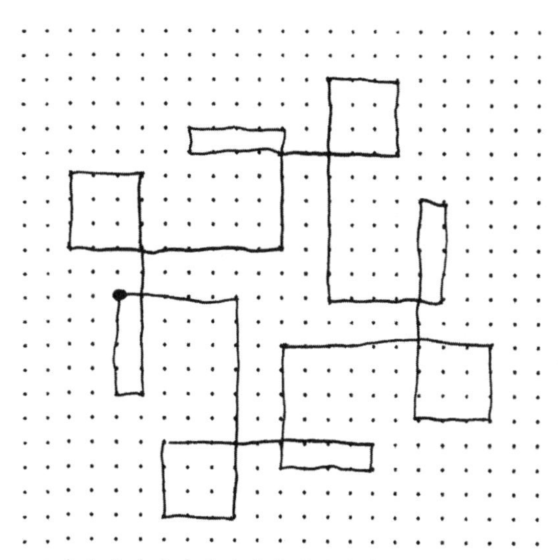

These are some students' reactions when asked to write about the hardest part of the activity. Clearly, there were students who had trouble mastering the process at first.

Remebering to turn the right way.

It was hard to count the dots.

This is about the hardest math lesson I have had ever done. But my friends and I pulled through in the lesson.

I rate this a 1000

Spirolaterals

Encourage your students to write about their experiences drawing spirolaterals and the impressions they have of their completed images. This kind of **communication** meets a variety of different goals.

MATH and FROGS

ROTATIONS

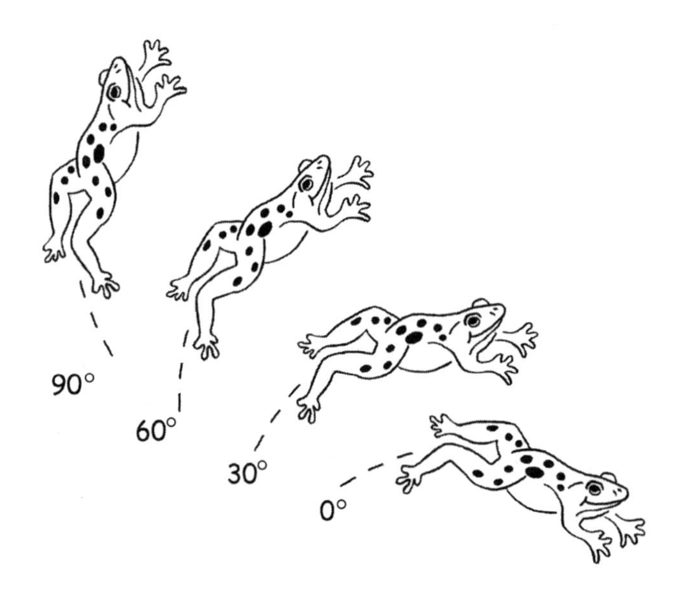

Introduction to Activity

PUTTING IT TOGETHER

Seeing Slides and Turns

Description

Cut a square picture of a frog into four square quarters. Then see how many ways the pieces can be reassembled back into the original square frame. Since there are only a limited number of discrete options for location and for orientation, use counting techniques from discrete math.

Students first count all of the different possible ways that the four pieces can be arranged in a square frame. Then other questions are explored. How can the original frog be rebuilt from some random arrangements of the four pieces, and how would you describe the process in terms of slides and turns of the cut quarters?

Start with these 4 pieces.

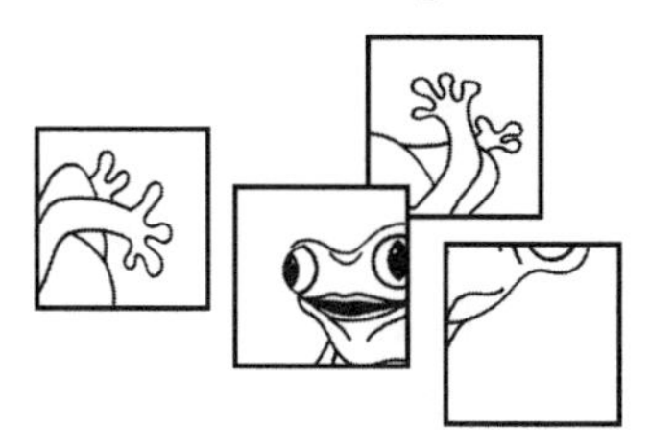

Put the head here.

There are 3 choices left for the next piece, with 3 possible locations and 4 ways it can be turned in the chosen location.

Big Idea

There are two key moves that are needed to reassemble a frog from the four square pieces. These transformations are **translations** and **rotations**. Just translate or slide the pieces into their correct locations in the frame and then rotate or turn them into their correct orientations.

Expected Outcomes

Students will gain experience in

- seeing separate parts of figures
- sliding and turning
- counting arrangements
- writing out steps
- visualizing transformations

Meeting these NCTM Standards

✓ Numbers	✓ Problem Solving
Algebra	✓ Reasoning and Proof
✓ Geometry	✓ Communication
✓ Measurement	✓ Connections
Data Analysis	✓ Representation

Activity 8 PUTTING IT TOGETHER

The Frog

Start with a frog in a square.

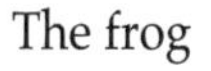

The frog

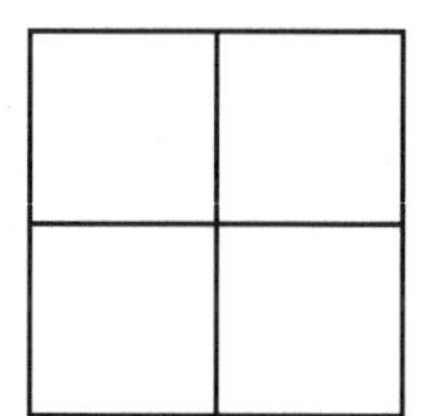

The frame

The Frame

Trace around the square to form a frame. Divide the frame into four smaller squares.

The Pieces

Cut the square frog picture in half, top to bottom, and left to right. Four smaller square parts are formed.

The cuts

The pieces

The Process

Now take the frame and the four pieces of the frog. What are some of the different ways that the pieces can be put back into the frame?

Here are some possible ways to **translate** and **rotate** the head piece into a specific location and orientation. There are many possible slides and turns that can be made.

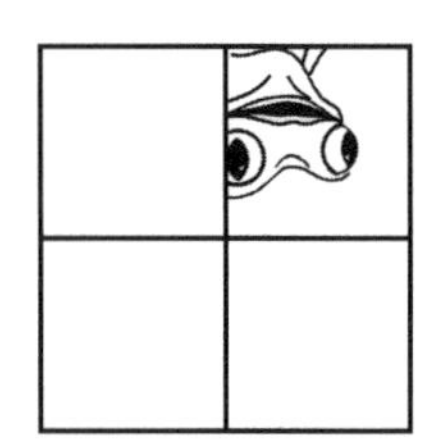
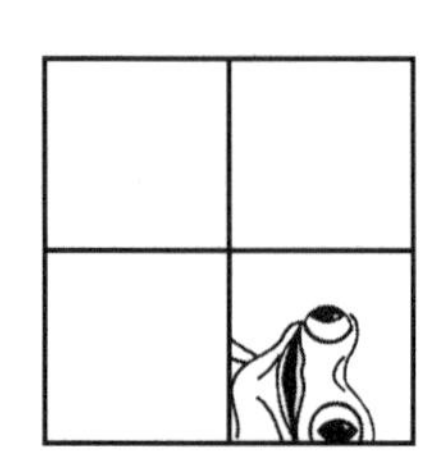

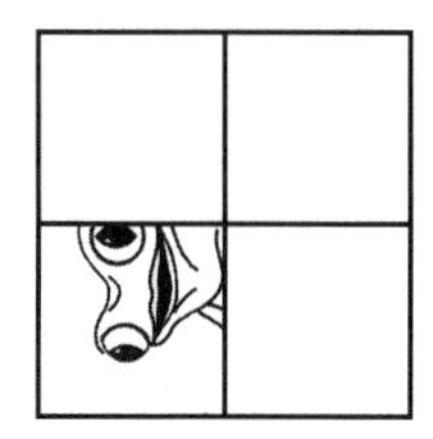

Here are some ways the other pieces might be added.

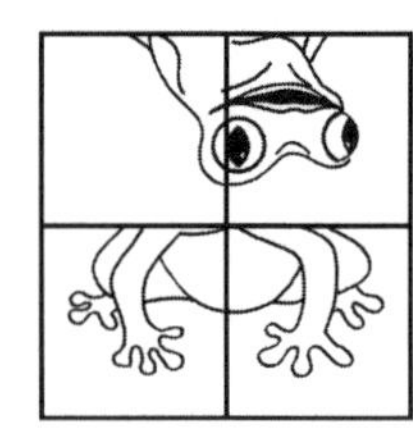
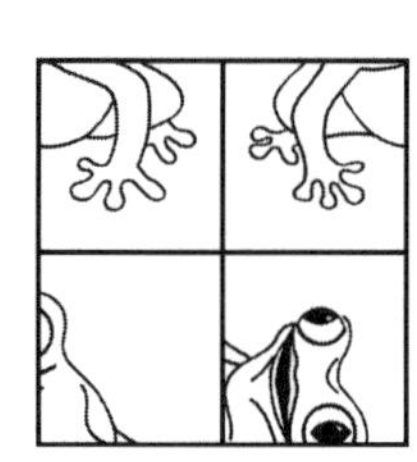
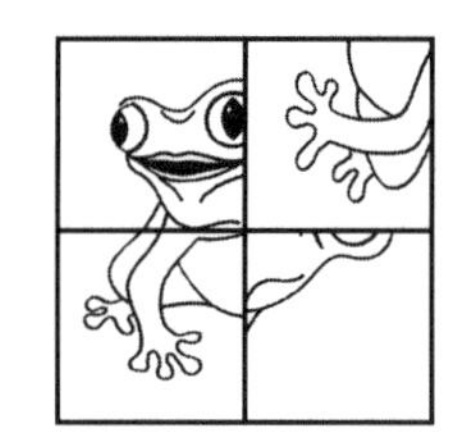
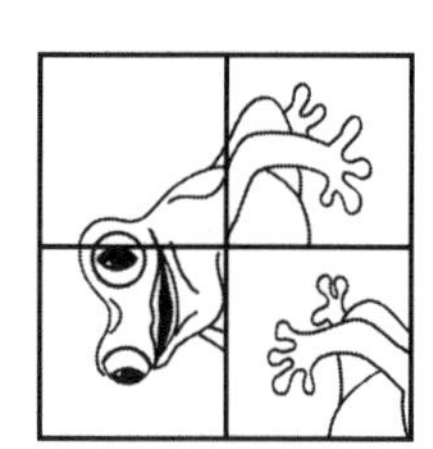

The Problem

How many different ways can the 4 square pieces be put back into the 4 different parts of the frame?

The Solution

The first piece can go in any one of the 4 parts of the frame. That leaves 3 choices for the second piece, 2 for the third, and 1 for the fourth. So there are 24 choices in all.

$$4 \times 3 \times 2 \times 1 = \mathbf{24}$$

But for each position in the frame that is chosen, 4 different turns are possible. In all, that makes 256 different turning arrangements possible for each choice of positions.

$$4 \times 4 \times 4 \times 4 = \mathbf{256}$$

The total number of possible arrangements of the four pieces in the frame is 6,144. Only one of these ways rebuilds the frog correctly and in the right position.

$$24 \times 256 = \mathbf{6{,}144}$$

The Counting Property

The key idea used here is the ***multiplication property for counting:***

> *If one event can occur in* **m** *ways and another in* **n** *ways,*
> *then the two together can occur in* **m × n** *ways.*

The Extension

Have students rebuild the frog using these **representations** for slides and turns.

Use these codes for ***location***:	**TL**	Top Left	**BL**	Bottom Left
They are the 4 translations.	**TR**	Top Right	**BR**	Bottom Right
Use these codes for ***rotation***:	**0**	No turn	**1/4**	One-quarter turn
They are the clockwise turns.	**1/2**	One-half turn	**3/4**	Three-quarters turn

0

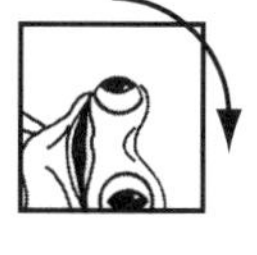

1/4

1/2

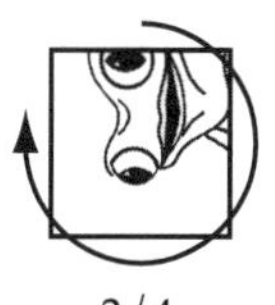

3/4

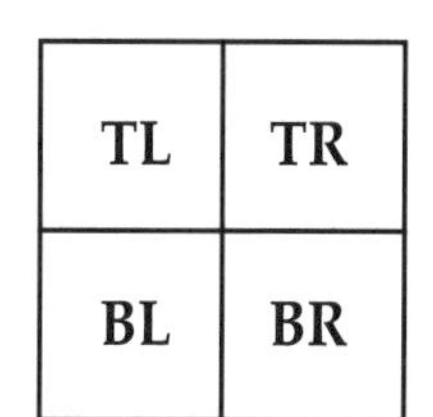

Lay out the four pieces. Then let the students do the **problem solving** needed to find the coded entries that tell where and how to move each piece to rebuild the frog in its correct position.

Slide to these locations and make these rotations to rebuild the frog correctly.

BR
1/4

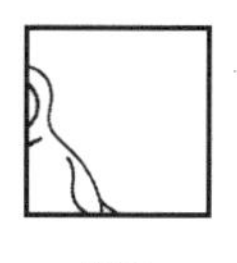

TR
0

TL
3/4

BL
1/2

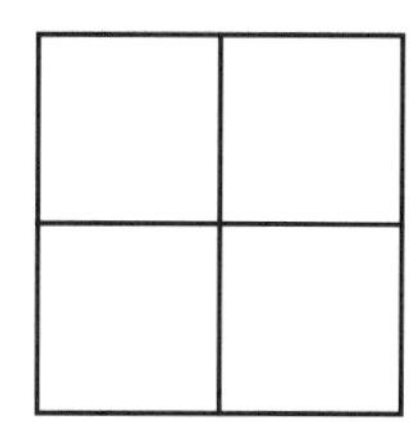

For another approach, have students choose and build any of the 6,144 possible arrangements. Then let their partners solve the puzzles by giving the translation and rotation codes needed for each piece in order to rebuild the frogs correctly.

COUNTING POSITIONS

Worksheet 8A

Circle the correct answer.

The frame The piece

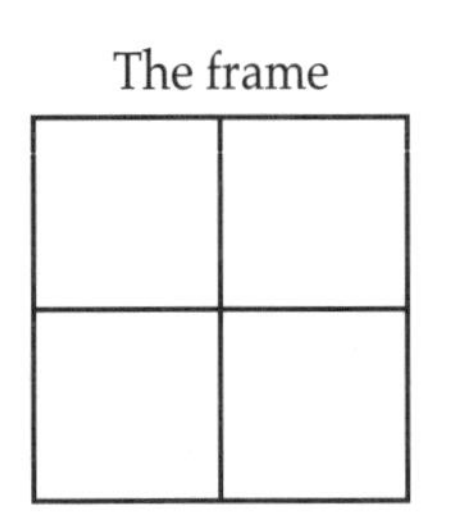

1. Slide the piece into the frame.
   How many locations are possible?

   1 2 3 4

2. Rotate the piece in a chosen location.
   How many turns are possible?

   1 2 3 4

3. How many choices of location and rotation are possible for this piece?

   1 x 4 2 x 4 3 x 4 4 x 4

The frame The piece

4. Slide this new piece into the frame.
   How many locations are possible?

   1 2 3 4

5. Rotate this piece in a chosen location.
   How many turns are possible?

   1 2 3 4

6. How many choices of location and rotation are possible for this piece?

   1 x 4 2 x 4 3 x 4 4 x 4

The frame The piece

7. Slide this new piece into this frame.
   How many locations are possible?

   1 2 3 4

8. Rotate this piece in a chosen location.
   How many turns are possible?

   1 2 3 4

9. How many choices of location and rotation are possible for this piece?

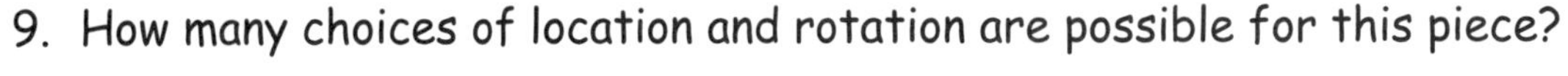

   1 x 4 2 x 4 3 x 4 4 x 4

REBUILDING THE FROG **Worksheet 8B**

The 4 pieces are lettered A, B, C, and D.
The 4 positions are named BL, TL, TR, and BR.

1. To rebuild the frog, slide piece A into position TR.
   Match the other pieces to their correct positions.

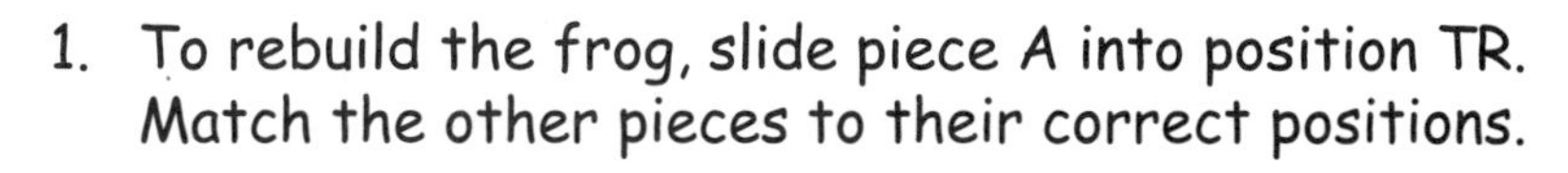

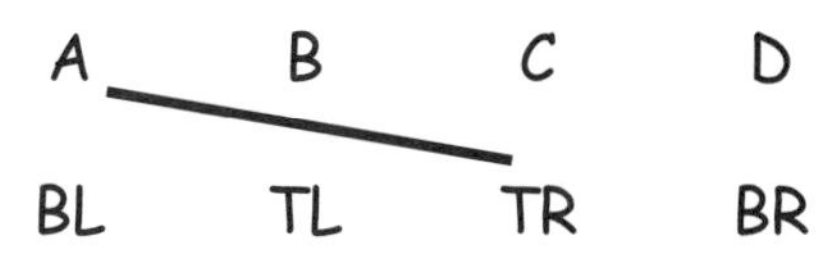

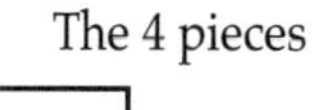
The 4 pieces

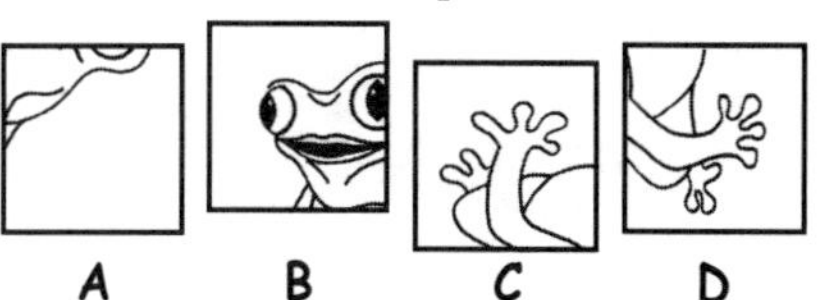

2. To rebuild the frog, piece D needs a 1/4 turn.
   Match the other pieces to their correct turns.
   Make all turns to the right, clockwise.

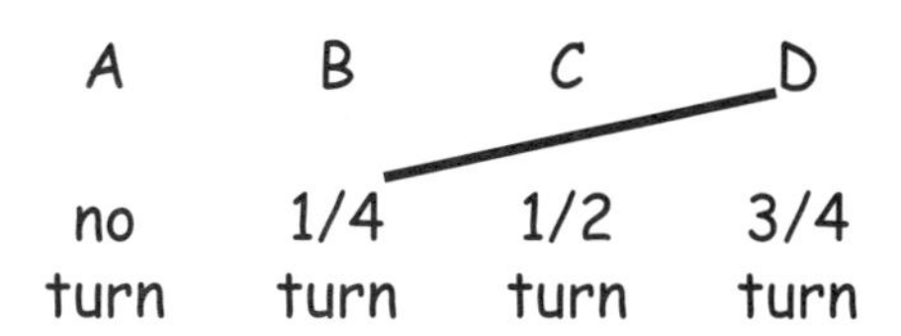

The 4 positions

TL	TR
BL	BR

3. Choose any position and any turn for piece A.
   Circle the number of choices that are possible?

   4 x 1 4 x 2 4 x 3 4 x 4

Complete these turns and slides to rebuild this frog in its correct form.

4. Rotate piece in TR a _____ turn to the right.
5. Rotate piece in BR a _____ turn to the right.
6. Slide piece in TL to position _____ .
7. Slide piece in TR to position _____ .
8. Translate piece in BL to position _____ .
9. Translate piece in BR to position _____ .

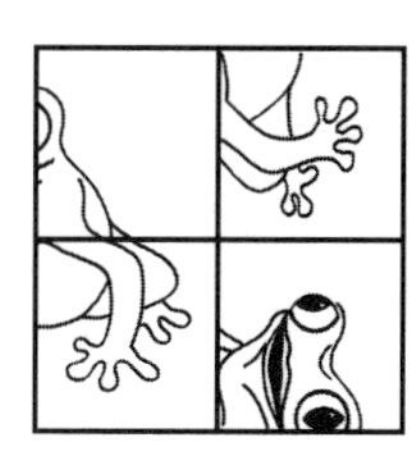

ANSWERS to Worksheets 8A and 8B

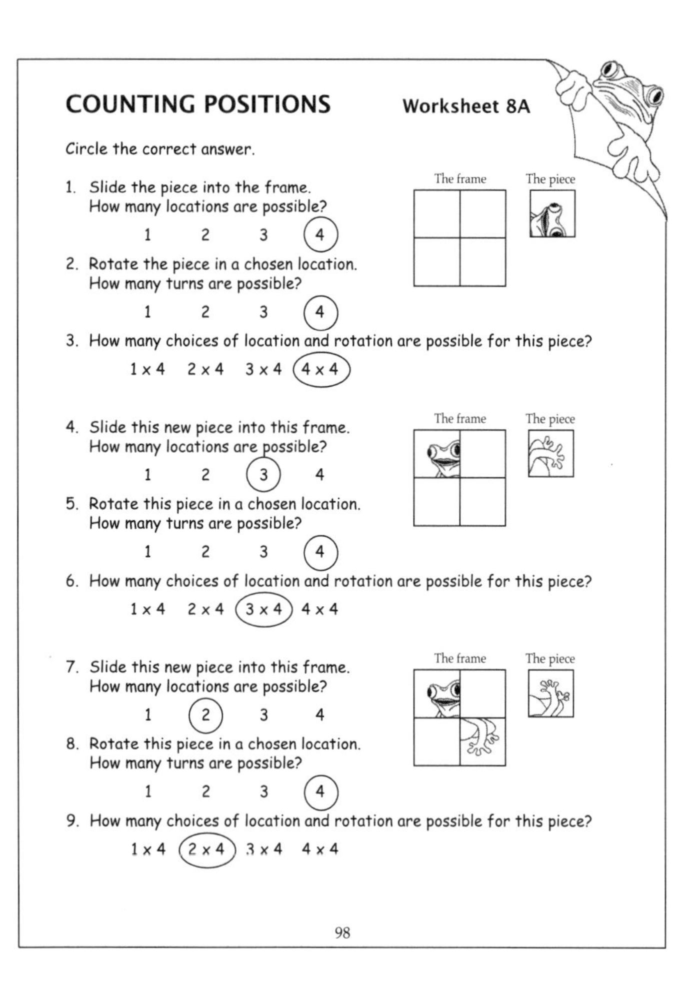

COUNTING POSITIONS **Worksheet 8A**

Circle the correct answer.

The frame The piece

1. Slide the piece into the frame.
   How many locations are possible?
   1 2 3 (4)
2. Rotate the piece in a chosen location.
   How many turns are possible?
   1 2 3 (4)
3. How many choices of location and rotation are possible for this piece?
   1 x 4 2 x 4 3 x 4 (4 x 4)

The frame The piece

4. Slide this new piece into this frame.
   How many locations are possible?
   1 2 (3) 4
5. Rotate this piece in a chosen location.
   How many turns are possible?
   1 2 3 (4)
6. How many choices of location and rotation are possible for this piece?
   1 x 4 2 x 4 (3 x 4) 4 x 4

The frame The piece

7. Slide this new piece into this frame.
   How many locations are possible?
   1 (2) 3 4
8. Rotate this piece in a chosen location.
   How many turns are possible?
   1 2 3 (4)
9. How many choices of location and rotation are possible for this piece?
   1 x 4 (2 x 4) 3 x 4 4 x 4

98

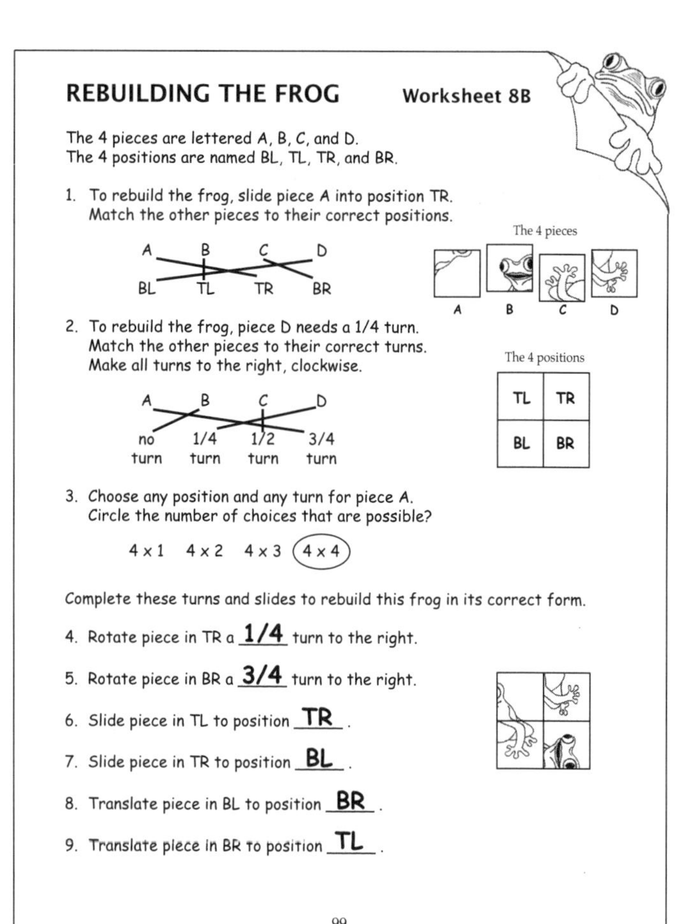

REBUILDING THE FROG **Worksheet 8B**

The 4 pieces are lettered A, B, C, and D.
The 4 positions are named BL, TL, TR, and BR.

1. To rebuild the frog, slide piece A into position TR.
   Match the other pieces to their correct positions.
   A B C D
   BL TL TR BR

The 4 pieces
A B C D

2. To rebuild the frog, piece D needs a 1/4 turn.
   Match the other pieces to their correct turns.
   Make all turns to the right, clockwise.
   A B C D
   no turn 1/4 turn 1/2 turn 3/4 turn

The 4 positions

TL	TR
BL	BR

3. Choose any position and any turn for piece A.
   Circle the number of choices that are possible?
   4 x 1 4 x 2 4 x 3 (4 x 4)

Complete these turns and slides to rebuild this frog in its correct form.

4. Rotate piece in TR a **1/4** turn to the right.
5. Rotate piece in BR a **3/4** turn to the right.
6. Slide piece in TL to position **TR**.
7. Slide piece in TR to position **BL**.
8. Translate piece in BL to position **BR**.
9. Translate piece in BR to position **TL**.

99

FROG Facts

A spawn, clutch, or cluster of female eggs may contain any number from the single egg of a tiny Cuban frog to 50 or so for the red-eyed tree frog to as many as 35,000 for the cane toad. The Eastern spadefoot toad lays its eggs in long, connected strings rather than clusters. The mortality rate for egg clusters may be as high as 95%.

TRANSPARENCY MASTERS

Make transparencies from these masters.

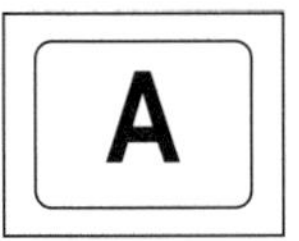

Cut out one frog intact. Cut the other two into quarters. Use these pieces and the empty frames from transparency B to build some of the many possible random arrangements of the four quarters.

These empty frames can be used in counting the different choices of both location and rotation. Slide the four cut quarters, one at a time, into the frame. Then show how each quarter can be turned in four different ways.

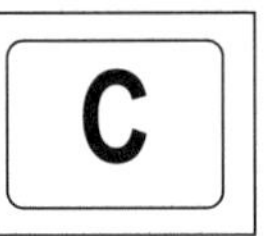

Together, these 32 arrangements show all possible choices for the two pieces on the right side, given that the other two on the left are already placed correctly. Note that three pieces are the same in each row of four and in each column of four.

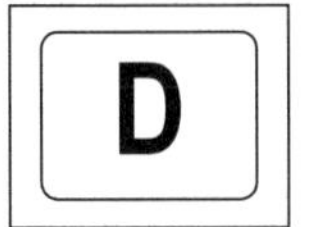

With this transparency, students can practice the visualization of translations and rotations as described in the suggested extension for this activity. Assume clockwise turns for the rotations. Where appropriate, encourage students to also give rotation answers in terms of degrees.

Mask out the solutions at the bottom at first. Use them later for students to check their work.

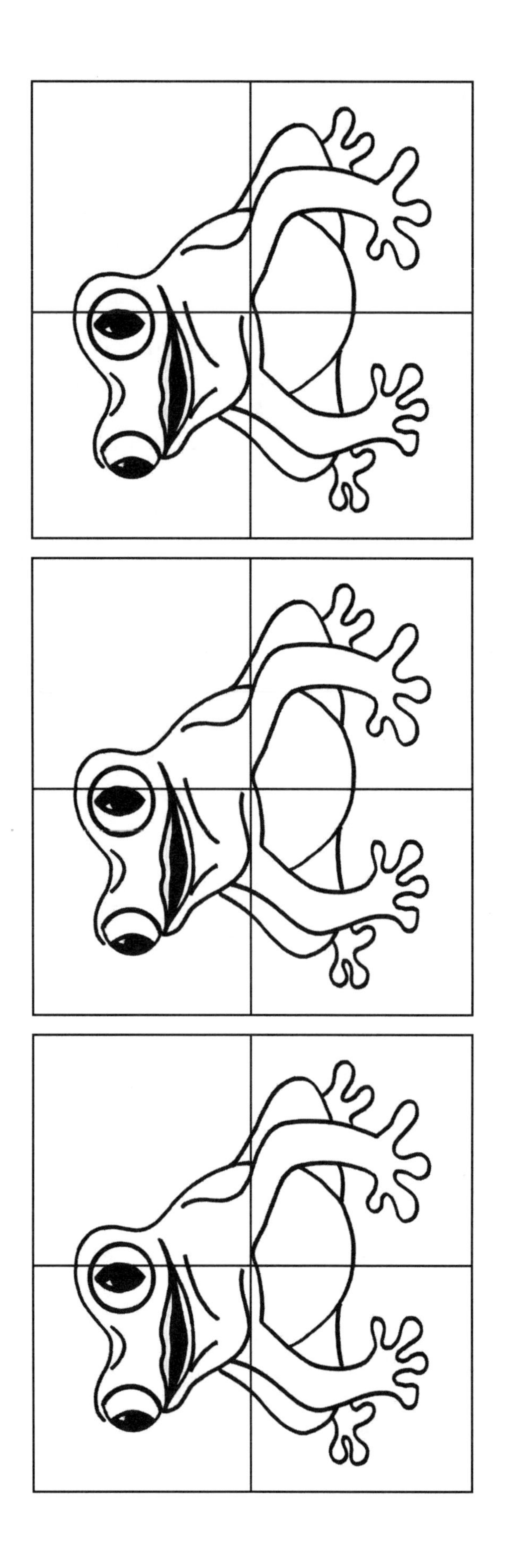

PUTTING THE PIECES TOGETHER

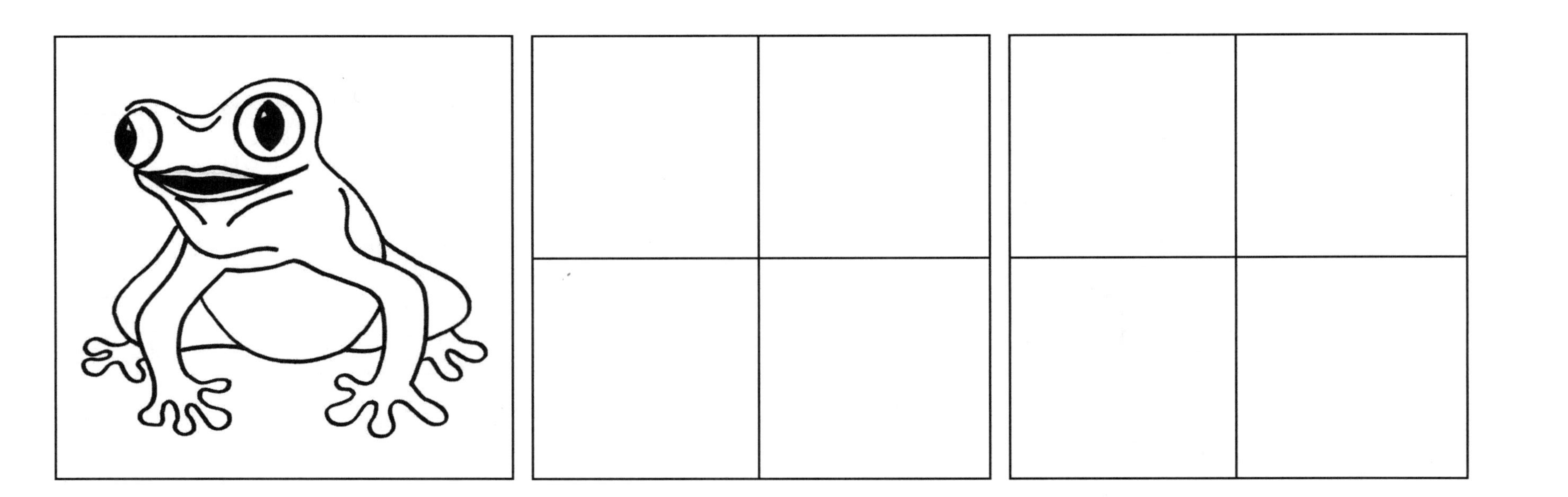

B

32 ARRANGEMENTS

with the two pieces on the left correctly positioned

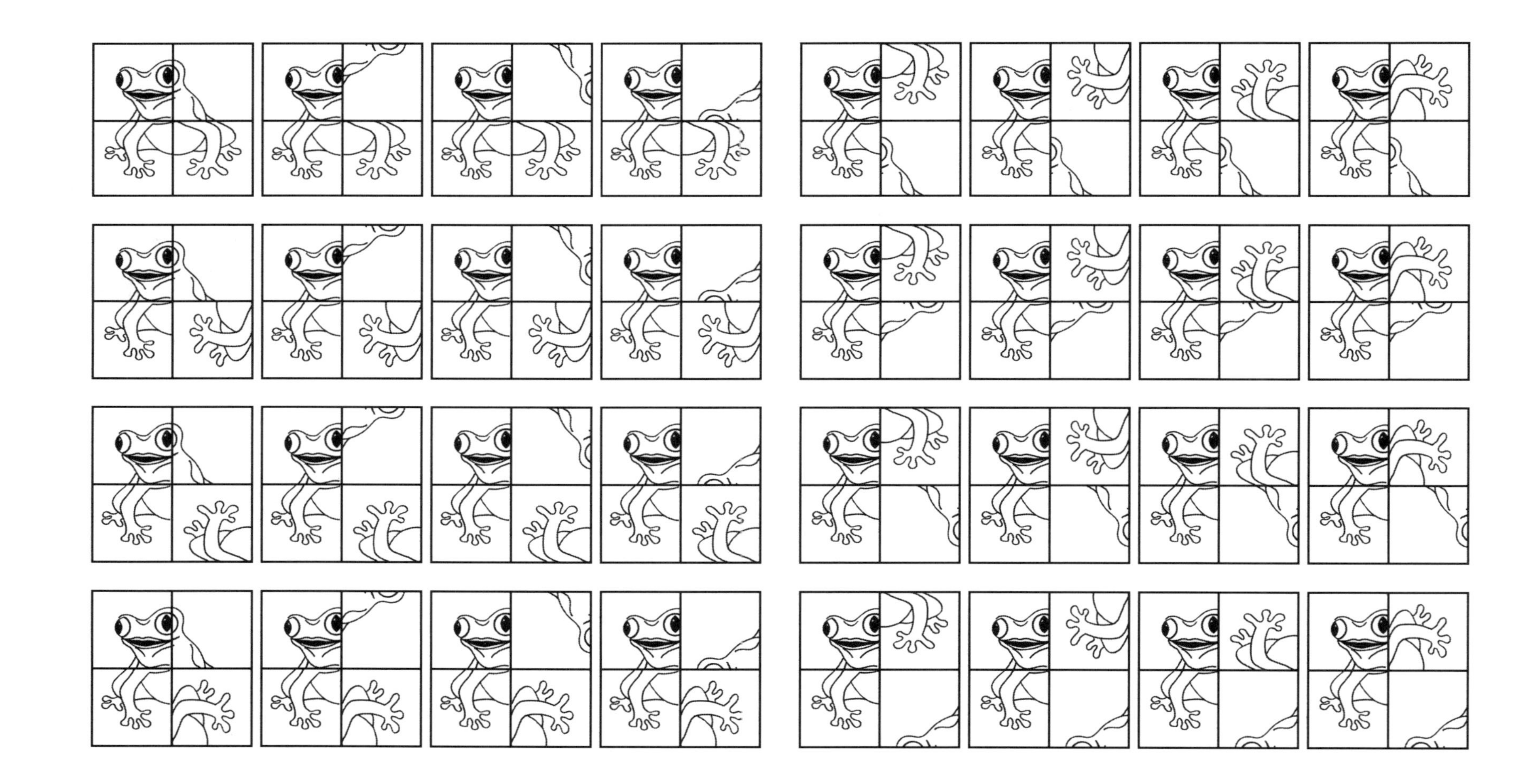

C

PUTTING IT TOGETHER

Rebuild each frog in its correct position.
Tell where and how to move each of the 4 pieces.

Use these codes.

TL	top left	**0**	no turn
TR	top right	**1/4**	one-quarter turn
BL	bottom left	**1/2**	one-half turn
BR	bottom right	**3/4**	three-quarter turn

1.

No slide Noturn	No slide 1/2 turn
No slide Noturn	No slide No turn

2.

Slide to BR No turn	Slide to BL No turn
Slide to TR No turn	Slide to TL No turn

3.

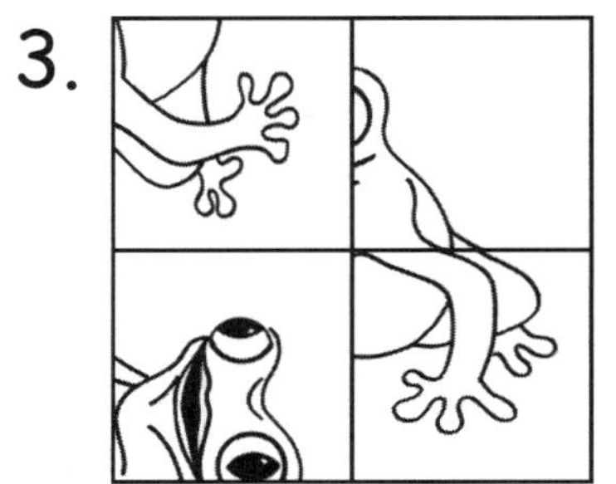

Slide to BL 1/4 turn	No slide No turn
Slide to TL 3/4 turn	No slide No turn

4.

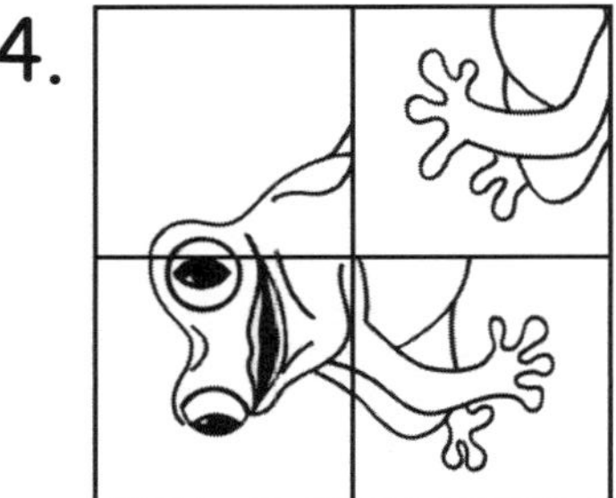

Slide to TR 1/4 turn	Slide to BR 3/4 turn
Slide to TL 1/4 turn	Slide to BL 1/4 turn

D

GETTING THE MOST from Activity 8

Use the empty frame on transparency B and the cut quarter pieces of acetate to show the steps in the counting process on the overhead projector. Focus on the **translation** and **rotation** transformations as you create different arrangements of the 4 pieces in the frame. Of the 6,144 possible choices, only one builds the frog in the correct position.

For a simpler problem, start with the head already in place. Count locations and rotations for the 3 remaining pieces.

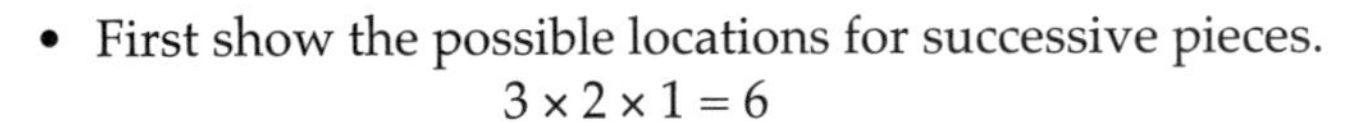

- First show the possible locations for successive pieces.
  $3 \times 2 \times 1 = 6$

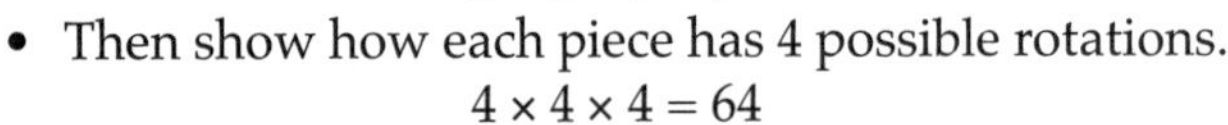

- Then show how each piece has 4 possible rotations.
  $4 \times 4 \times 4 = 64$

- Finally, multiply to count locations and rotations together.

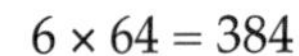

  $6 \times 64 = 384$

Place 3 pieces

If 2 pieces are already correctly placed, the total number of possible arrangements is reduced to 32.

- First show the possible locations for successive pieces,
  $2 \times 1 = 2$
- Then show how each piece has 4 possible rotations.
  $4 \times 4 = 16$
- Finally, multiply to count locations and rotations together.
  $2 \times 16 = 32$

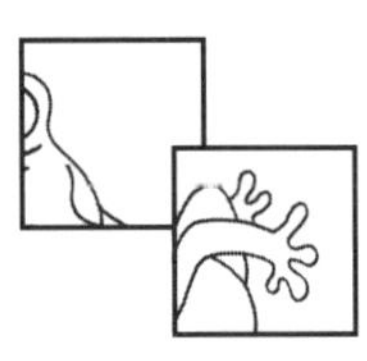

Place 2 pieces

This last set of 32 arrangements would make an interesting bulletin board display, with each arrangement pasted together by a student. A transparency master is included showing all of these 32 different possibilities.

As a suggested extension activity that involves the skill of **communication**, have individuals paste up their own random arrangements of the 4 pieces. Then let them exchange their figures and see if their partners can correctly describe in writing how to rebuild the frogs.

- Expect a variety of methods to be used. Some students will slide their pieces first. Others will turn them first. As noted by this student, having a second frog, uncut, might be helpful.

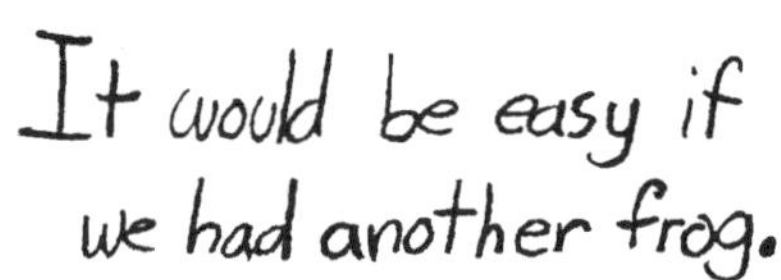

- Right and left hand turns are determined from the top, the right being clockwise and the left counter-clockwise. Note that this student's error would turn the frog's head upside down.

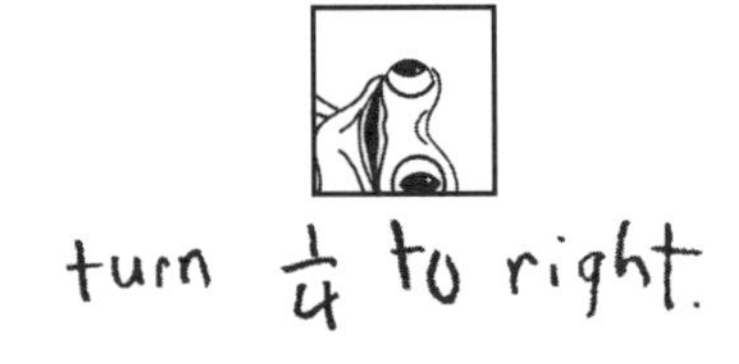

STUDENT REACTIONS to Activity 8

Check out how Jennifer wrote the solution to her cut-up frog puzzle.

In this case, the choice was made to turn all four pieces into the correct orientations before sliding any of them into their correct locations.

TL TR

BL BR

For a more advanced, more symbolic **representation** of the same rebuilding steps, describe the slides and turns in the form shown below. Each entry tells where and how to move that piece.

Slide to TR 1/4 turn left	Slide to BL 1/2 turn right
Slide to TL 1/2 turn right	Slide to BR 1/4 turn left

1. TL turn 1 quarter left.
2. BL half turn right
3. BR 1 quarter left turn
4. turn TR half turn right
5. Slide TL to TR Slide TR to BL
6. BL slide to TL

Jennifer

Students were asked to make suggestions that might help other students who would be doing this **problem-solving** activity later. Some of their remarks are quite interesting.

It is easier to rotate before sliding.

As these comments show, there was disagreement on whether sliding or turning should be done first.

I t's easyer to slide before you rotate.

I think it is eser to turn first then slide the peasas first.

rotate befor glue.

Yes, it would be harder to rotate the pieces after they are glued in place.

MATH and FROGS

FRACTIONS

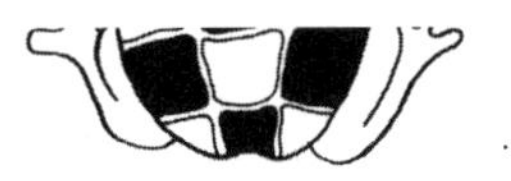

What fractional part of this frog is missing?
Is 1/2, 1/3, or 1/4 the best estimate?

PUZZLES for PROBLEM SOLVING

PUZZLE A ALPHABET PATTERNS 110

Alphabetizing with a twist

Students create an alphabetized list of 2- and 3-letter arrangements of the letters in FROG. Then they count those that are meaningful English words.

PUZZLE B YEARS PER INCH 112

Spanning time on a line

This activity has students see time through the eyes of length using a time line. It also reviews computational skills and fundamental number line concepts.

PUZZLE C MORE CHOICES OF BUGS 113

Creating a new lunch menu

A frog eats three bugs for lunch and has three different kinds to choose from. Each 3-letter branch on the tree diagram represents one of the 27 choices.

PUZZLE D LETTERING CUBES 114

Changing 2-D figures into 3-D solids

Students complete the lettering on 2-dimensional nets by visualizing how they would appear when cut out and folded into 3-dimensional cubes.

PUZZLE E STRANGE ANIMALS 115

Repeatingpatternspatternspatterns...

Solve these spirolateral puzzles by finding the repeating geometric patterns. Then convert them back into number patterns, letters, and finally words.

PUZZLE F LEAPFROG COUNTING 116

Letters on both sides

Letter a paper strip of squares with L, E, A, P on one side and F, R, O, G on the other. Count all possible four-letter arrangements of the separate squares.

PUZZLE G COORDINATES 117

Using number pairs to follow the moves

Ordered number pairs are used in this puzzle where pieces of the frog are arranged and rearranged and their positions recorded using coordinates.

ANSWERS 118

ALPHABET PATTERNS

Puzzle A

Imagine a single, combined, alphabetized list of all possible arrangements of one, two, three, and four of the letters in the word FROG. The list would be quite long.

Here are the first four entries in alphabetical order.

F
FG
FGO
FGOR

1. What are the fifth, sixth, and seventh entries in an alphabetized list of arrangements? Be careful. Each one will have a different number of letters in it.

2. The eighth entry is a meaningful English word. What is it?

3. Make a complete list, in alphabetical order, of the 16 different one-, two-, three-, and four-letter arrangements that start with F. Remember, no letter can be repeated in any single arrangement.

4. If an arrangement is chosen at random from the list above, what are the chances that it is a meaningful English word?

5. What is the total number of all possible arrangements of one-, two-, three-, and four-letter words starting with any of the four letter? Think carefully. See if you can find a quick way to get the answer.

6. Here are the 64 possible arrangements of one, two, three, or four of the letters in the word FROG. They have been alphabetized into four sets of 16, one set starting with each of the four letters F, R, O, and G. Describe the repeating pattern in the number of letters in each of these four alphabetized sets.

F	G	O	R
FG	GF	OF	RF
FGO	GFO	OFG	RFG
FGOR	GFOR	OFGR	RFGO
FGR	GFR	OFR	RFO
FGRO	GFRO	OFRG	RFOG
FO	GO	OG	RG
FOG	GOF	OGF	RGF
FOGR	GOFR	OGFR	RGFO
FOR	GOR	OGR	RGO
FORG	GORF	OGRF	RGOF
FR	GR	OR	RO
FRG	GRF	ORF	ROF
FRGO	GRFO	ORFG	ROFG
FRO	GRO	ORG	ROG
FROG	GROF	ORGF	ROGF

7. How many of these 64 different arrangements of 1, 2, 3, or 4 letters are meaningful English words?

8. If one arrangement is chosen at random from the complete list above, what are the chances that it is a meaningful English word?

YEARS PER INCH

Puzzle B

Seconds, minutes, hours, days, weeks, months, and years are all different units that measure the same thing, time. Length also can be measured in many different units, including inches, feet, yards and miles. Different units of time can be compared to one another, as can different units of length.

60	minutes per hour	12	inches per foot
24	hours per day	3	feet per yard
365	days per year	5280	feet per mile

1. How many minutes are in one day?
   How many minutes are in one year? ______ ______

2. How many inches are in one yard? ______ ______
   How many inches are in one mile?

On a time line, time in years can be compared to length in inches. There are 2000 years of time represented by the 8-inch length of a paper strip.

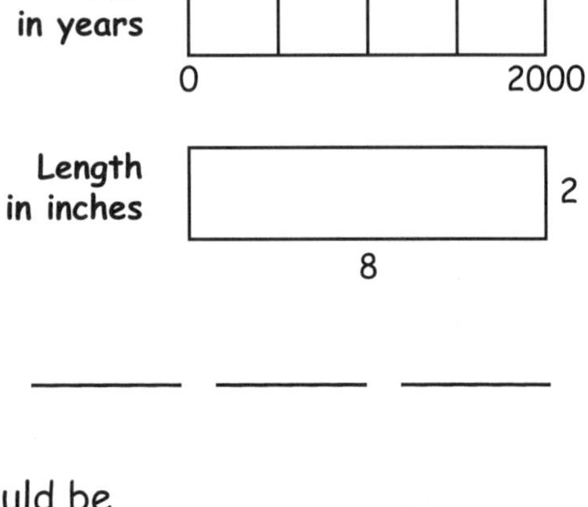

3. If there are 2000 years per 8 inches, how many years are there per 4 inches? per 2 inches? per inch? ______ ______ ______

4. At 250 years per inch, how many inches would be needed for a time line of one million years? ______

5. The first frog fossils date back 150 million years. At 250 years per inch, would a 1-mile paper strip be long enough to reach back 150 million years? If not, how many miles long would the strip have to be?

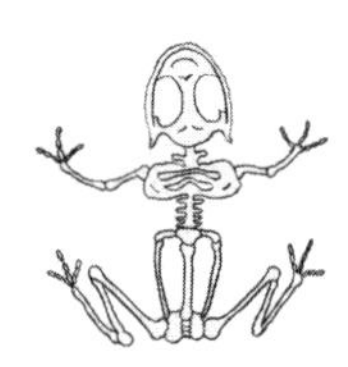

MORE CHOICES OF BUGS **Puzzle C**

A frog catches three bugs for lunch. Each bug is a fly, a gnat, or a mosquito. This tree diagram shows all the different possible choices for the three-bug lunch.

f is for fly.
g is for gnat.
m is for mosquito.

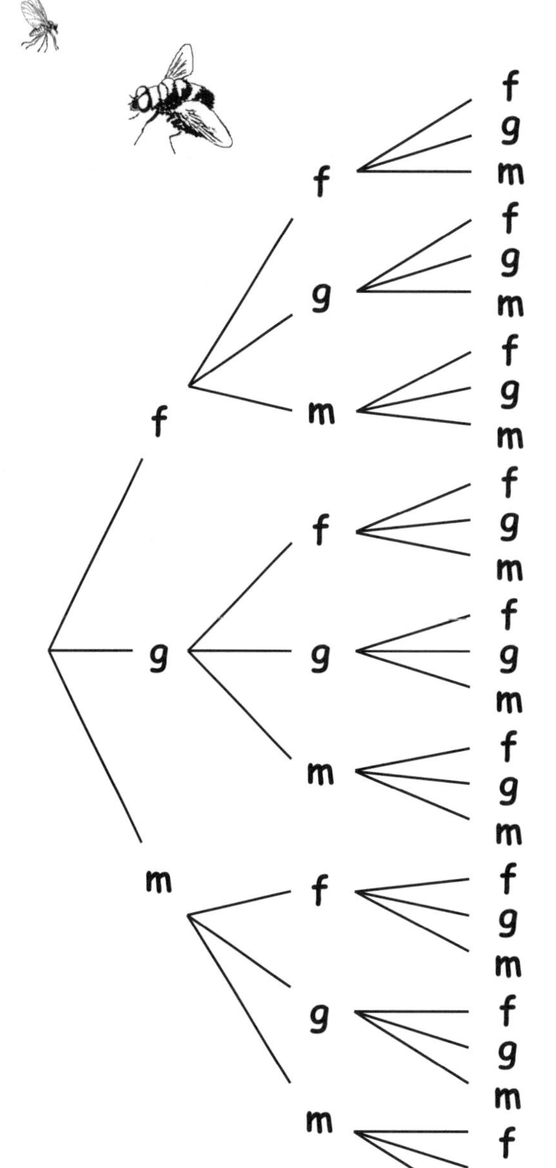

1. How many choices of a three-bug lunch are possible? _____

2. How many choices have all the bugs alike? _____

What are the chances that a three-bug lunch has

3. all the bugs alike?

   _____ out of _____

4. just two bugs alike?

   _____ out of _____

Express the probability for each three-bug lunch as a fraction.

5. all three bugs are gnats _____

6. just two bugs are gnats _____

7. only one bug is a gnat _____

8. no bug is a gnat _____

LETTERING CUBES

Puzzle D

These six squares form a 2-D net for a cube. When cut out and folded up, they form the six faces of a 3-D cube that spells the word FROG.

Place the missing letters for the word FROG in the nets below. Choose the correct square and rotation for each letter. When built, the cube should spell the word FROG just as the one on the right would.

F	R	O	G

1. F R O

2. R F

3. F R

4. F

5. Now cut out each net. Fold it up into a cube. Check to see if you placed the letters in the correct squares, turned the correct ways.

STRANGE ANIMALS

Puzzle E

You've seen the spirolateral for the word FROGS. Here it is again with the code.

1	A J S
2	B K T
3	C L U
4	D M V
5	E N W
6	F O X
7	G P Y
8	H Q Z
9	I R

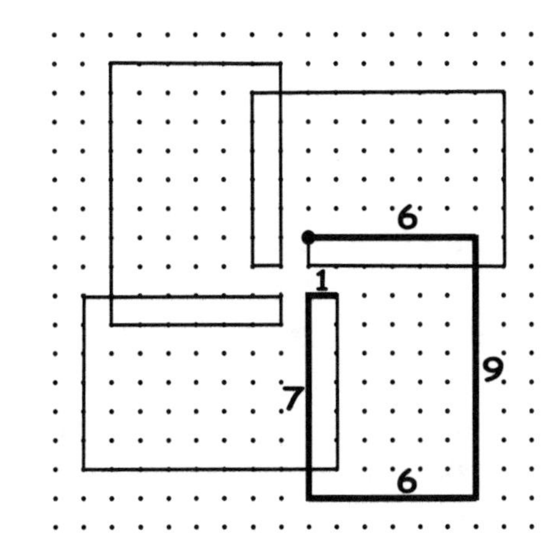

FROGS
69671

Frogs and toads live around other animals. See if you can figure out these animal names from their spirolaterals. Look for the repeating number sequences first. Then see if you can figure out the names from the numbers.

1.

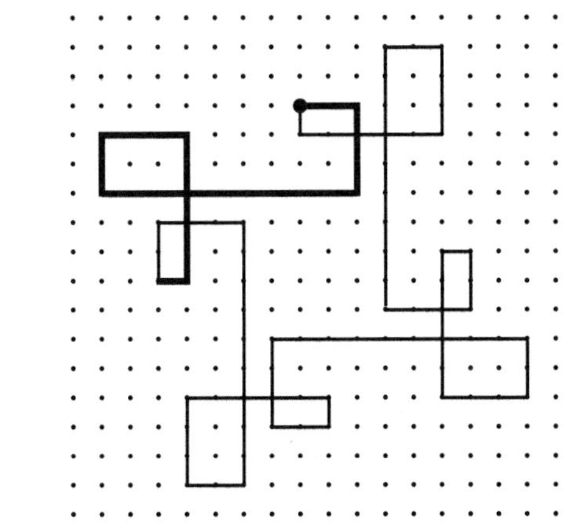

2.

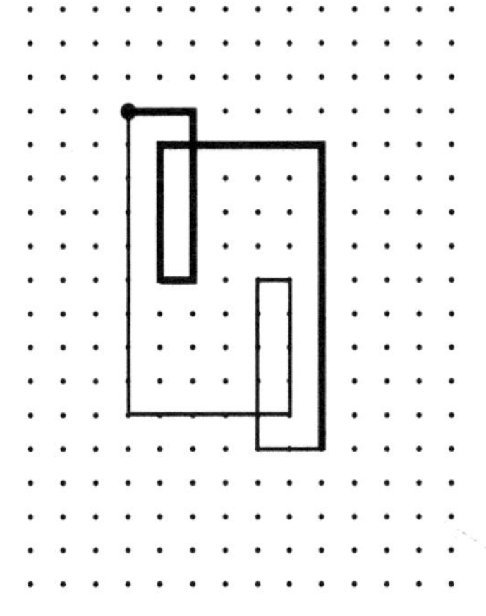

3.

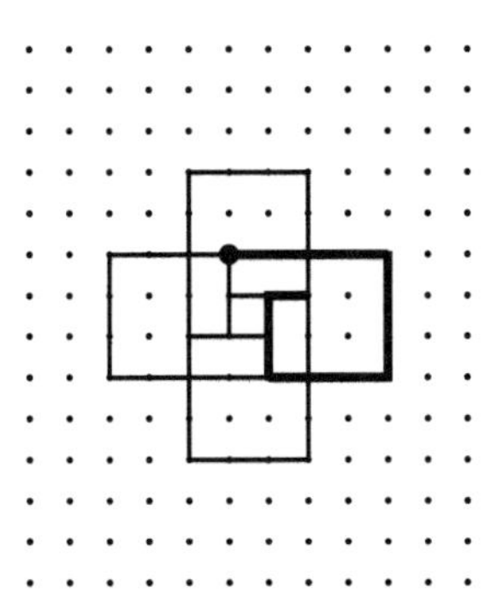

4.

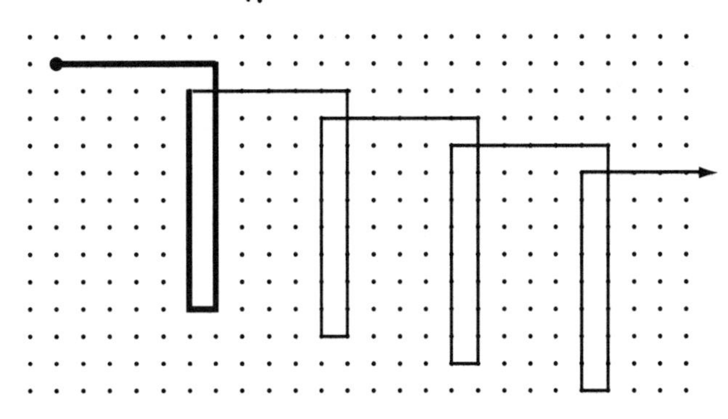

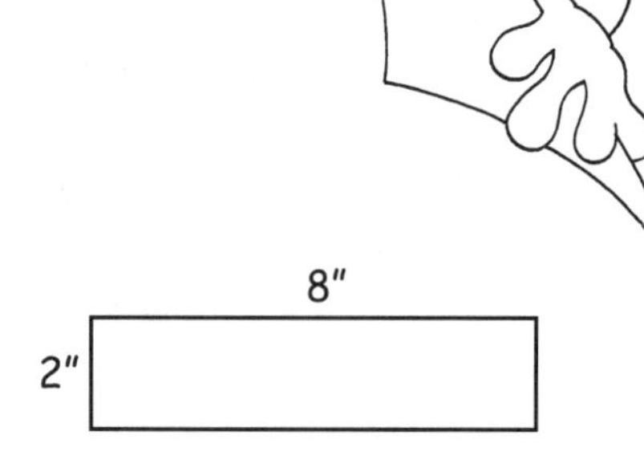

LEAPFROG COUNTING

Puzzle F

Start with a 2 x 8-inch strip of paper. Fold it in half and in half again. Then unfold the strip.

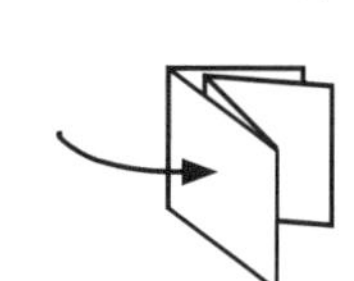

Mark the four squares on one side with the letters L, E, A, and P. On the other side put an F behind the P, an R behind the A, an O behind the E, and a G behind the L.

Tear the strip into the four separate squares. There are eight letters, but you only have four paper squares to move around and arrange.

Can you form these four-letter words? Answer yes or no for each one.

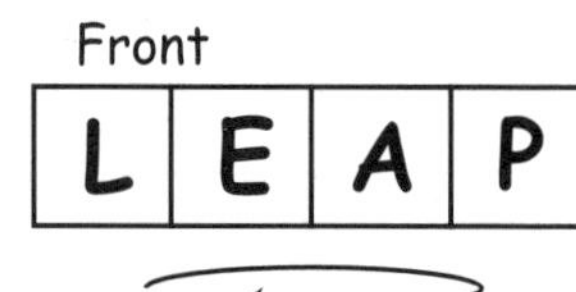

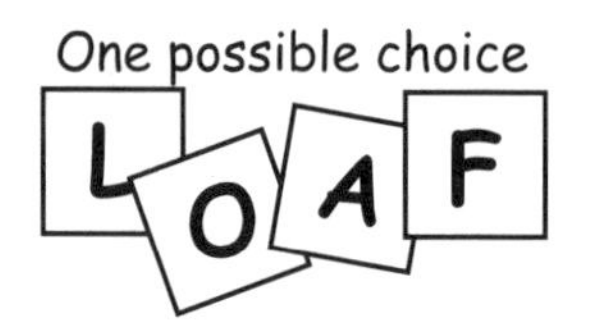

1. PALE ______ 2. POLE ______
3. LOAF ______ 4. LEAF ______
5. GAPE ______ 6. GOAL ______
7. ROPE ______ 8. FLAG ______
9. OPAL ______ 10. FOAL ______

Imagine an alphabetized list of all the many ways to arrange these four paper squares in a row,

11. What would be the first entry in the list? What would be the last? ______ ______

12. Show how you would find the total number of possible arrangements in the entire list.

COORDINATES

Puzzle G

Coordinates are ordered number pairs. They give the locations of points as measured distances, first along the x-axis and then along the y-axis. Here are the coordinates for point A on the upper right square. It is a vertex on one of the square pieces of this frog that needs rebuilding.

vertex **A**
on the upper right square **(2 , 2)**

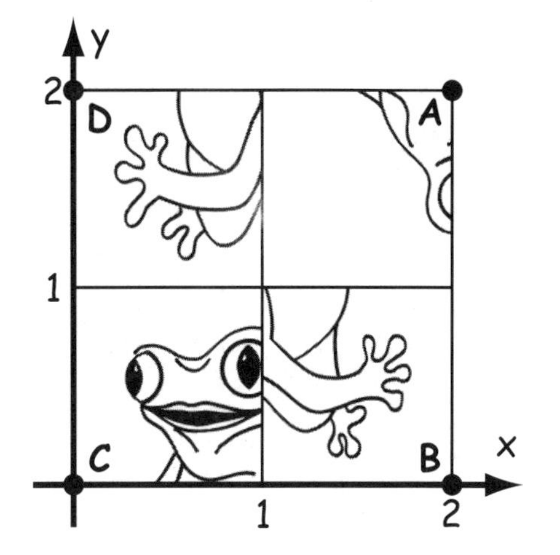

Give the coordinates for these vertices.

1. vertex **B**
   on the lower right square (__ , __)

2. vertex **C**
   on the lower left square (__ , __)

3. vertex **D**
   on the upper left square (__ , __)

Now imagine these same four square pieces rearranged so as to rebuild the frog in its correct shape and position. Vertex A, on the upper right piece, will move from (2, 2) to (1, 1). We can represent this change this way.

vertex **A** **(2 , 2) → (1 , 1)**

In the same fashion, show where these vertices will go when the four pieces are rearranged to rebuild the frog in its correct shape and position.

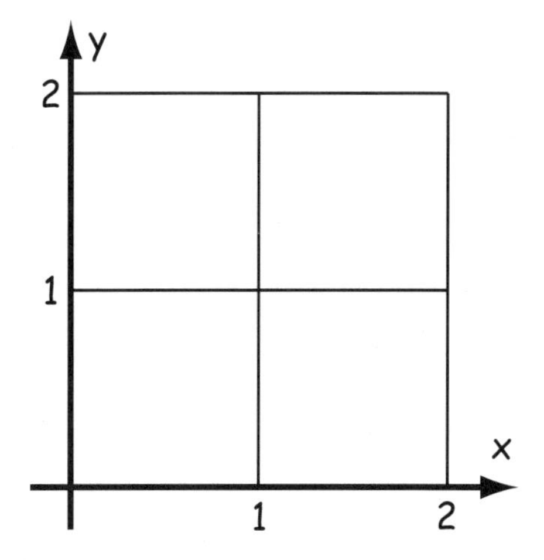

5. vertex **B** (__ , __) → (__ , __)

6. vertex **C** (__ , __) → (__ , __)

7. vertex **D** (__ , __) → (__ , __)

8. Plot the new location of points **A**, **B**, **C**, and **D** on the grid shown.

ANSWERS to Puzzles A, B, and C

ALPHABET PATTERNS — Puzzle A

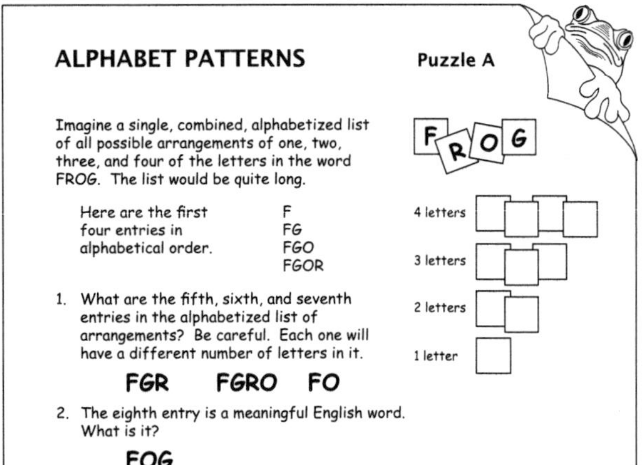

Imagine a single, combined, alphabetized list of all possible arrangements of one, two, three, and four of the letters in the word FROG. The list would be quite long.

Here are the first four entries in alphabetical order.
F
FG
FGO
FGOR

1. What are the fifth, sixth, and seventh entries in the alphabetized list of arrangements? Be careful. Each one will have a different number of letters in it.

   FGR FGRO FO

2. The eighth entry is a meaningful English word. What is it?

   FOG

3. Make a complete list, in alphabetical order, of the 16 different one-, two-, three-, and four-letter arrangements that start with F. Remember, no letter can be repeated in any single arrangement.

F	**FGR**	**FOGR**	**FRG**
FG	**FGRO**	**FOR**	**FRGO**
FGO	**FO**	**FORG**	**FRO**
FGOR	**FOG**	**FR**	**FROG**

4. If an arrangement is chosen at random from the list above, what are the chances that it is a meaningful English word?

   4 chances out of 16 FOG, FOR, FRO, FROG

5. What is the total number of all possible arrangements of one-, two-, three-, and four-letter words starting with any of the four letter? Think carefully. See if you can find a quick way to get the answer.

   4 x 16 = 64

110

6. Here are the 64 possible arrangements of one, two, three, or four of the letters in the word FROG. They have been alphabetized into four sets of 16, one set starting with each of the four letters F, R, O, and G. Describe the repeating pattern in the number of letters in each of these four alphabetized sets.

F	G	O	R	**1**
FG	GF	OF	RF	**2**
FGO	GFO	OFG	RFG	**3**
FGOR	GFOR	OFGR	RFGO	**4**
FGR	GFR	OFR	RFO	**3**
FGRO	GFRO	OFRG	RFOG	**4**
FO	GO	OG	RG	**2**
FOG	GOF	OGF	RGF	**3**
FOGR	GOFR	OGFR	RGFO	**4**
FOR	GOR	OGR	RGO	**3**
FORG	GORF	OGRF	RGOF	**4**
FR	GR	OR	RO	**2**
FRG	GRF	ORF	ROF	**3**
FRGO	GRFO	ORFG	ROFG	**4**
FRO	GRO	ORG	ROG	**3**
FROG	GROF	ORGF	ROGF	**4**

7. How many of these 64 different arrangements of 1, 2, 3, or 4 letters are meaningful English words?

   8 FOG, FOR, FRO, FROG, GO, O, OF, OR

8. If one arrangement is chosen at random from the complete list above, what are the chances that it is a meaningful English word?

   8 chances out of 64

111

YEARS PER INCH — Puzzle B

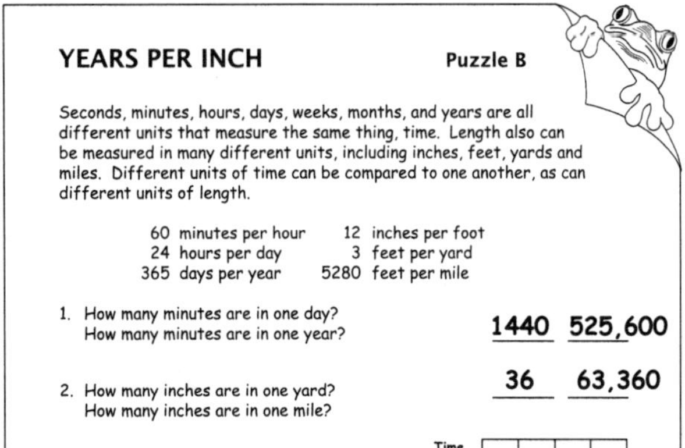

Seconds, minutes, hours, days, weeks, months, and years are all different units that measure the same thing, time. Length also can be measured in many different units, including inches, feet, yards and miles. Different units of time can be compared to one another, as can different units of length.

60 minutes per hour
24 hours per day
365 days per year

12 inches per foot
3 feet per yard
5280 feet per mile

1. How many minutes are in one day? How many minutes are in one year? **1440** **525,600**

2. How many inches are in one yard? How many inches are in one mile? **36** **63,360**

On a time line, time in years can be compared to length in inches. There are 2000 years of time represented by the 8-inch length of a paper strip.

3. If there are 2000 years per 8 inches, how many years are there per 4 inches? per 2 inches? per inch? **1000** **500** **250**

4. At 250 years per inch, how many inches would be needed for a time line of one million years? **4000**

5. The first frog fossils date back 150 million years. At 250 years per inch, will a 1-mile paper strip be long enough to reach back 150 million years? If not, about how many miles long will the strip have to be?

   No. **600,000 inches**
   50,000 feet
   about 9 1/2 miles

112

MORE CHOICES OF BUGS — Puzzle C

A frog catches three bugs for lunch. Each bug is a fly, a gnat, or a mosquito. This tree diagram shows all the different possible choices for the three-bug lunch.

f is for fly.
g is for gnat.
m is for mosquito.

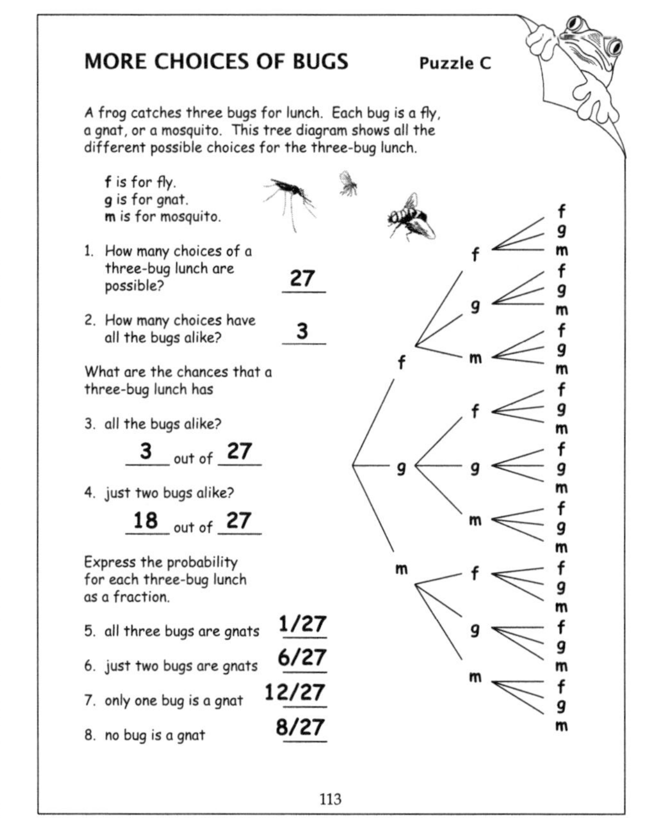

1. How many choices of a three-bug lunch are possible? **27**

2. How many choices have all the bugs alike? **3**

What are the chances that a three-bug lunch has

3. all the bugs alike? **3** out of **27**

4. just two bugs alike? **18** out of **27**

Express the probability for each three-bug lunch as a fraction.

5. all three bugs are gnats **1/27**

6. just two bugs are gnats **6/27**

7. only one bug is a gnat **12/27**

8. no bug is a gnat **8/27**

113

ANSWERS to Puzzles D, E, F, and G

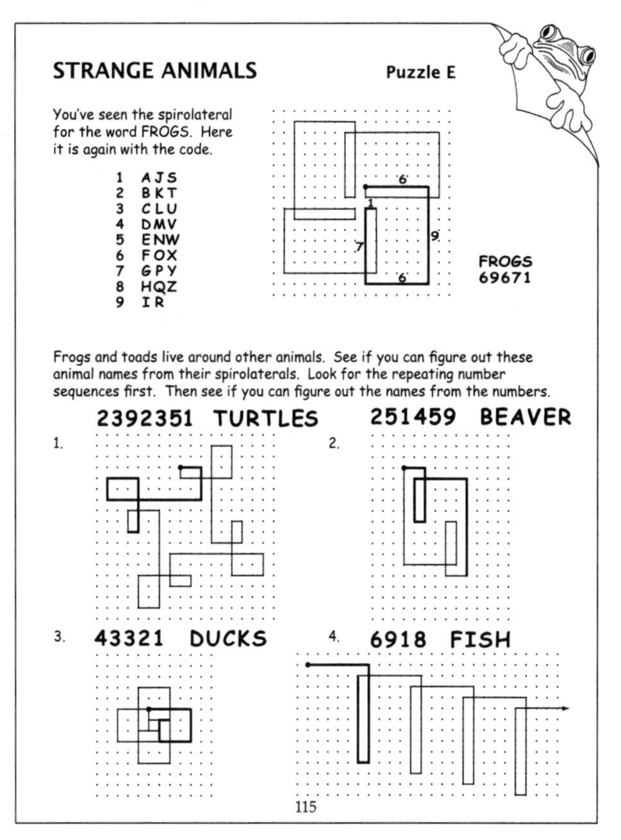

LEAPFROG COUNTING

Puzzle F

Start with a 2 x 8-inch strip of paper. Fold it in half and in half again. Then unfold the strip.

Mark the four squares on one side with the letters L, E, A, and P. On the other side put an F behind the P, an R behind the A, an O behind the E, and a G behind the L.

Tear the strip into the four separate squares. There are eight letters, but you only have four paper squares to move around and arrange.

Can you form these four-letter words? Answer yes or no for each one.

1. PALE **yes** 2. POLE **no**
3. LOAF **yes** 4. LEAF **yes**
5. GAPE **yes** 6. GOAL **no**
7. ROPE **no** 8. FLAG **no**
9. OPAL **yes** 10. FOAL **yes**

Front: L E A P
Back: F R O G
One possible choice: L O A F

Imagine an alphabetized list of all the many ways to arrange these four paper squares in a row,

11. What would be the first entry in the list? What would be the last? **AEFG RPOL**

12. Show how you would find the total number of possible arrangements in the entire list.

8x6x4x2 = 384

116

COORDINATES

Puzzle G

Coordinates are ordered number pairs. They give the locations of points as measured distances, first along the x-axis and then along the y-axis. Here are the coordinates for point A on the upper right square. It is a vertex on one of the square pieces of this frog that needs rebuilding.

vertex **A** on the upper right square **(2 , 2)**

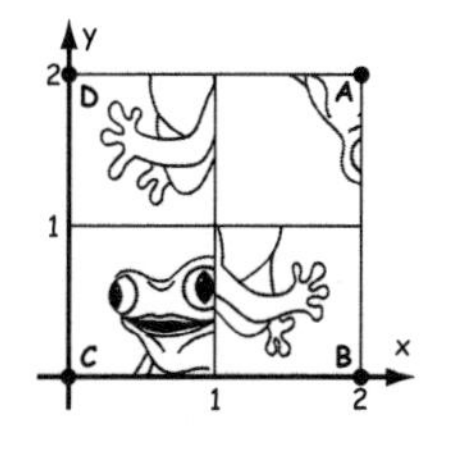

Give the coordinates for these vertices.

1. vertex **B** on the lower right square (**2** , **0**)
2. vertex **C** on the lower left square (**0** , **0**)
3. vertex **D** on the upper left square (**0** , **2**)

Now imagine these same four square pieces rearranged so as to rebuild the frog in its correct shape and position. Vertex A, on the upper right piece, will move from (2, 2) to (1, 1). We can represent this change this way.

vertex **A** (2 , 2) → (1 , 1)

In the same fashion, show where these vertices will go when the four pieces are rearranged to rebuild the frog in its correct shape and position.

5. vertex B (**2** , **0**) → (**0** , **0**)
6. vertex C (**0** , **0**) → (**0** , **1**)
7. vertex D (**0** , **2**) → (**1** , **0**)

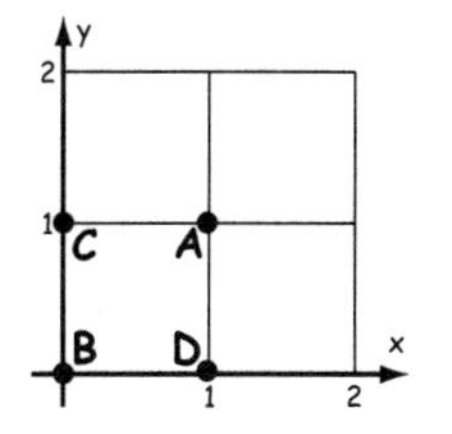

8. Plot the new location of points **A**, **B**, **C**, and **D** on the grid shown.

117

WRITING IDEAS

Students have an innate interest and curiosity in solving problems and puzzles and in reading about things that interest them. Capitalize on this, but at the same time challenge them to express their ideas, thoughts, and findings in words.

1. *Frogs have a very interesting life cycle that goes through several different transformations. Read and write about these stages in the life of a frog.*

2. *A time line is 5 inches long. The year 1900 is at the beginning and 2100 is at the end. Explain in words how to find where to mark the year 1985.*

3. *Draw your own net for a cube. Be creative in arranging the squares. Then explain how you would letter four faces to spell FROG when assembled.*

4. *Read about red-eyed tree frogs. Then make up a math problem related to one of the frog facts that you found. Be certain to include the answer.*

5. *Write about how and where you would find the word FROG in a complete alphabetized list of all 24 different arrangements of the letters in the word.*

6. *Explain in your own words the number pattern used to write additional rows in the triangular array of numbers called Pascal's triangle.*

7. *Make a time line locating the dates of some of the important events in your life thus far. Then write a brief description of each of the dates you marked.*

8. *Read the story called* ***The Frog Prince****, written by the Grimm Brothers, and write a brief summary of the story.*

READING IDEAS

There are many interesting books at this level that connect literature to mathematics. Likewise, there are numerous books and internet sites that can serve as valuable references on frogs. Search them out and read them yourself. Then encourage your students to read some as well.

Literature Connections

Anno, Masaichiro & Mitsumasa. *Anno's Mysterious Multiplying Jar.* New York: Philomel Books, 1983.

Scieszka, Jon. *Math Curse.* New York: Viking, 1995.

Science Connections

Badger, David. *Frogs.* Stillwater, MN: Voyageur Press, 1995.

Berger, Melvin & Gilda. *How Do Frogs Swallow With Their Eyes?* New York: Scholastic, Inc., 2002.

Nature Conservancy. *Frogs & Toads.* New York: Houghton Mifflin Co., 1992.

Mathematics Connections

Kohl, Herbert. *Mathematical Puzzlements: Play & Invention with Mathematics.* New York: Schocken Books, 1987.

Sobel, Max & Evan Maletsky. *Teaching Mathematics: A Sourcebook of Aids, Activities, and Strategies.* Boston, MA: Allyn & Bacon, 1999.

Internet Connections

Exploratorium: *Frogs.* http://www.exploratorium.edu/frogs/links.html

Hamline University Center for Global Environmental Education: http://cgee.hamline.edu/frogs/

STRETCHING

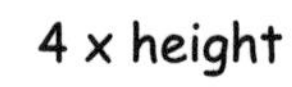

4 x height   2 x height   normal